ADAPTIVE RESONANCE THEORY MICROCHIPS

Circuit Design Techniques

THE KLUWER INTERNATIONAL SERIES IN ENGINEERING AND COMPUTER SCIENCE

ADAPTIVE RESONANCE THEORY MICROCHIPS

TERESA SERRANO-GOTARREDONA
BERNABÉ LINARES-BARRANCO
National Microelectronics Center, Sevilla, Spain

ANDREAS G. ANDREOU
The Johns Hopkins University

KLUWER ACADEMIC PUBLISHERS
Boston/London/Dordrecht

Distributors for North, Central and South America:
Kluwer Academic Publishers
101 Philip Drive
Assinippi Park
Norwell, Massachusetts 02061 USA

Distributors for all other countries:
Kluwer Academic Publishers
Distribution Centre
Post Office Box 322
3300 AH Dordrecht, THE NETHERLANDS

Library of Congress Cataloging-in-Publication Data

A C.I.P. Catalogue record for this book is available
from the Library of Congress.

Copyright © 1998 by Kluwer Academic Publishers

All rights reserved. No part of this publication may be reproduced, stored in a retrieval system or transmitted in any form or by any means, mechanical, photocopying, recording, or otherwise, without the prior written permission of the publisher, Kluwer Academic Publishers, 101 Philip Drive, Assinippi Park, Norwell, Massachusetts 02061

Printed on acid-free paper.

Printed in the United States of America

Contents

List of Figures	ix
List of Tables	xvi
Preface	xvii

1. ADAPTIVE RESONANCE THEORY ALGORITHMS ... 1
   1.1 Introduction ... 1
   1.2 ART1 ... 2
   1.3 ARTMAP ... 14
   1.4 Fuzzy-ART ... 23
   1.5 Fuzzy-ARTMAP ... 30

2. A VLSI-FRIENDLY ART1 ALGORITHM ... 39
   2.1 The Modified ART1 Algorithm ... 40
   2.2 Functional Differences between Original and Modified Model ... 43

3. ART1 AND ARTMAP VLSI CIRCUIT IMPLEMENTATION ... 53
   3.1 Introduction ... 53
   3.2 Hardware-Oriented Attractive Properties of the ART1 Algorithm ... 55
   3.3 Circuit Description ... 57
   3.4 Modular System Expansivity ... 63
   3.5 Implementation of Synaptic Current Sources ... 64
   3.6 Experimental Results of First Prototype ... 67
   3.7 Experimental Results of Second Prototype ... 76

4. A CURRENT-MODE MULTI-CHIP WTA-MAX CIRCUIT ... 89
   4.1 Introduction ... 89
   4.2 Operation Principle ... 92
   4.3 Circuit Implementation ... 94
   4.4 System Stability Coarse Analysis ... 100
   4.5 System Stability Fine Analysis ... 105
   4.6 Experimental Results ... 109

5. AN ART1/ARTMAP/FUZZY-ART/FUZZY-ARTMAP CHIP 117
   5.1 The Synaptic Cell 118
   5.2 Peripheral Cells 128
   5.3 Concluding Remarks 143

6. ANALOG LEARNING FUZZY ART CHIPS 145
   6.1 Introduction 145
   6.2 Summary of the Fuzzy-ART Algorithm 146
   6.3 Current-Mode Fuzzy-ART Chip 147
   6.4 Fuzzy-ART/VQ Chip 154
   6.5 Conclusions 164

7. SOME POTENTIAL APPLICATIONS FOR ART MICROCHIPS 167
   7.1 Portable Non-invasive Device for Determination of Concentrations of Biological Substances 169
   7.2 Cardiac Arrhythmia Classifier for Implantable Pacemaker 172
   7.3 Vehicle Interior Monitoring Device for Auto Alarm 174
   7.4 Concluding Remarks 175

Appendices 177
A– MATLAB Codes for Adaptive Resonance Theory Algorithms 177
   A.1 MATLAB Code Example for ART1 177
   A.2 MATLAB Code Example for ARTMAP 179
   A.3 MATLAB Code Example for Fuzzy-ART 181
   A.4 MATLAB Code Example for Fuzzy-ARTMAP 183
   A.5 Auxiliary Functions 185
B– Computational Equivalence of the Original ART1 and the Modified ART1m Models 191
   B.1 Direct Access to Subset and Superset Patterns 193
   B.2 Direct Access by Perfectly Learned Patterns (Theorem 1 of original ART1) 194
   B.3 Stable Choices in STM (Theorem 2 of original ART1) 196
   B.4 Initial Filter Values determine Search Order (Theorem 3 of original ART1) 196
   B.5 Learning on a Single Trial (Theorem 4 of original ART1) 197
   B.6 Stable Category Learning (Theorem 5 of original ART1) 198
   B.7 Direct Access after Learning Self-Stabilizes (Theorem 6 of original ART1) 199
   B.8 Search Order (Theorem 7 of original ART1) 202
   B.9 Biasing the Network towards Uncommitted Nodes 206
   B.10 Expanding Proofs to Fuzzy-ART 206
   B.11 Remarks 207
C– Systematic Width-and-Length Dependent CMOS Transistor Mismatch Characterization 209
   C.1 Mismatch Characterization Chip 210

C.2	Mismatch Parameter Extraction and Statistical Characterization	213
C.3	Characterization Results	215

References 223

Index 233

List of Figures

1.1	Topological structure of the ART1 architecture	3
1.2	Algorithmic description of ART1 functionality	5
1.3	Training sequence of ART1 with randomly generated input patterns	8
1.4	Illustration of the self-scaling property. (a) One mismatch out of three features, (b) one mismatch out of fifteen features	10
1.5	Simple ARTMAP architecture	15
1.6	Different exemplars for letters 'A', 'B', and 'C'	16
1.7	Flow diagram for simple ARTMAP architecture learning operation mode	17
1.8	Flow diagram for simple ARTMAP architecture recall operation mode	19
1.9	General ARTMAP architecture	22
1.10	Topological structure of the Fuzzy-ART architecture	24
1.11	Flow Diagram of the Fuzzy-ART Operation	25
1.12	Expansion of the rectangle R_j when a new input vector **b** is incorporated into category j	28
1.13	Set of Points used for Training of Fuzzy-ART	30
1.14	Categories formed by a Fuzzy-ART system trained with the training set shown in Fig. 1.13 with parameters $\alpha = 0.1$ and (a) $\rho = 0.3$, (b) $\rho = 0.6$, (c) $\rho = 0.7$, and (d) $\rho = 0.8$.	31
1.15	Flow Diagram for Fuzzy-ARTMAP Architecture	32
1.16	The two spirals used to train Fuzzy-ARTMAP	33
1.17	Test set results of Fuzzy-ARTMAP for the two spiral problem for different $\overline{\rho_a}$ values and $\alpha = 0.001$. The training set are the spiral points and the test set is the 100×100 grid points	35
1.18	Flow Diagram of Fuzzy-ARTMAP Algorithm in Slow-Learning Mode	37

2.1	Different algorithmic implementations of the ART1 architecture: (a) original ART1 (b) ART1 with a single binary valued weight template (c) and VLSI-friendly ART1m	41
2.2	Illustration of the simplification process of the division operation: (a) original division operation, (b) piece-wise linear approximation, (c) linear approximation	42
2.3	Comparative Learning Example ($\rho = 0.4, L = 5, \alpha = 2$)	46
2.4	Simulated Results Comparing Behavior between ART1 and ART1m	48
2.5	Learned Categories Average Distances	49
2.6	Optimal parameters fit between ART1 and ART1m	50
3.1	Hardware Block Diagram for the ART1m Algorithm	58
3.2	(a) Details of Synaptic Circuit S_{ij}, (b) Details of Controlled Current Source Circuit C_i	61
3.3	Circuit Schematic of Winner-Takes-All (WTA) Circuit	61
3.4	(a) Circuit Schematic of Current Comparator using Digital Inverter or (b) OTA, (c) of Active-Input Regulated-Cascode Current Mirror, (d) and of Adjustable Gain ρ Current Mirror.	62
3.5	Interchip Connectivity for Modular System Expansion	63
3.6	Cascade of Current Mirrors for Low Mismatch	65
3.7	Optimum Strategy for Minimum Current Sources Mismatch for Large Area Arrays	67
3.8	Measured Mismatch Error (in %) between 18 arbitrary L_A Current Sources	69
3.9	Set of Input Patterns	71
3.10	Clustering Sequence for $\rho = 0.7$ and $\alpha = L_A/L_B = 1.05$	72
3.11	Categorization of the Input Patterns for $\alpha = 1.07$ and different values for ρ	73
3.12	Categorization of the Input Patterns for $\rho = 0$ and different values of α	74
3.13	Categorization of the Input Patterns performed by Operative Samples	77
3.14	Measured Current for an Array of MOS Transistors with the same V_{GS} and V_{DS} voltages (for a nominal current of $10\mu A$), spread over a die area of $2.5mm \times 2.5mm$. (a) Array of NMOS Transistors, (b) Array of PMOS Transistors.	78
3.15	Measured Synaptic Current as a Function of Synapse Position in the Chip, for (a) L_{A1} Current Sources, (b) L_{A2} Current Sources, and (c) L_B Current Sources.	81
3.16	Training Sequence for a one-chip ART1 System with $\rho = 0.3$ and $\alpha = 1.1$	83
3.17	Training Sequence for a two-chip ART1 System with $\rho = 0.5$ and $\alpha = 2$	84

3.18	(a) ARTMAP Hardware Assembly, (b) Diagram of Inter-ART chip, (c) Detail of Inter-ART Chip Cell	85
3.19	Complete Training Sequence of the ARTMAP System for $\overline{\rho_a} = 0$ and $\rho_b = 0.75$	87
3.20	Recognition Sequence performed on a previously trained ARTMAP System. Applied Input Patterns are Noisy Versions of the Training Set.	88
4.1	WTA topologies. (a) WTA of $O(M^2)$ complexity, (b) transformation to $O(M)$ complexity, (c) typical topology of $O(M)$ WTA hardware implementation	91
4.2	Graphic Representation of the Solution of eq. (4.4)	93
4.3	WTA unit cell: (a) circuit diagram, (b) transfer curve	95
4.4	Diagram of the WTA circuit	95
4.5	Strategy to assemble several chips	96
4.6	Enhanced current mirror topologies, (a) active, (b) cascode, (c) regulated cascode output, and (d) active regulated cascode	98
4.7	Resolution (in bits) as a function of working current for different current mirror topologies	101
4.8	Small Signal Delay Modeling for N-output Current Mirror	103
4.9	(a) Parallel connection of unstable cells, (b) uncompensated small signal equivalent circuit, (c) compensated small signal equivalent circuit.	106
4.10	Transfer curves of the WTA implemented in a ES2 $1.0\mu m$ chip for an input current level of $100\mu A$ (horizontal scale is $3\mu A/div$ from 90 to $110\mu A$, vertical scale is $0.15V/div$ from $0V$ to $1.65V$).	111
4.11	Transfer curves when two ES2 $1.0\mu m$ chips are assembled and for an input current level of $10\mu A$ (horizontal scale is $0.2\mu A/div$ from 9 to $11\mu A$, vertical scale is $0.125V/div$ from $0V$ to $1.375V$).	111
4.12	Transfer curves when two chips of different technology are assembled and for an input current level of $500\mu A$ (horizontal scale is $10\mu A/div$ from $450\mu A$ to $550\mu A$, vertical scale is $0.375V/div$ from $0V$ to $4.125V$).	112
4.13	(a) Input Signals, (b) Output Waveforms	113
5.1	Conceptual Diagram of Current Source Flip-Flop	120
5.2	Grouping of Current-Source Flip-Flops to assemble an n-bit Current-Source Memory Cell suitable for Fuzzy-ART Synaptic Cells	121
5.3	(a) Physical Layout for CMOS Compatible Lateral Bipolar (pnp) Transistor, (b) Equivalent Schematic Representation of the resulting Parallel Connection of Bipolar pnp and PMOS.	122

5.4	(a) Current Source Flip-Flop with Bipolar pnp Transistors as Current Sources, (b) Physical Layout of a Two-Collector Lateral Bipolar pnp Structure.	123
5.5	Source/Emitter driven p-type current mirrors, (a) bipolar and (b) MOS versions	123
5.6	Conventional Active-Input Current Mirror, (a) bipolar and (b) MOS versions	124
5.7	Small Signal Equivalent Circuits for (a) Source Driven Active Input, and (b) Conventional Active Input Current Mirrors	125
5.8	Alternative Source Driven Active Input Current Mirror	125
5.9	Complete Circuit Schematic of ART1/Fuzzy-ART Synaptic Cell	127
5.10	Layout of ART1/Fuzzy-ART Synaptic Cell	129
5.11	Floorplan of ART1/Fuzzy-ART/ARTMAP/Fuzzy-ARTMAP Chip	129
5.12	Schematic Circuit for Bottom Cells	132
5.13	Simplified Circuit Diagram of Top Cells	134
5.14	Circuit Configuration of Right Cells for Fuzzy-ART Operation Mode with (a) subtraction based and (b) division based choice functions	136
5.15	Circuit Configuration of Right Cells for ART1 Operation Mode with (a) subtraction based and (b) division based choice functions	137
5.16	(a) Complete Schematic of Current Division Circuit, (b) Basic Processing Circuit	138
5.17	Circuit Diagram for Vigilance Circuit	140
5.18	One Branch of the WTA Circuit	141
5.19	Complete Layout of Small Size Prototype Test Chip	142
6.1	Chip architecture for $N = 4$ and $M = 8$	148
6.2	The learning cell	150
6.3	Evolution of weights for different β_{small}, $\beta = 0.05$, $\gamma = 2.4 \times 10^{-5}$	153
6.4	Measured Fuzzy-min Computation	154
6.5	Measured Normal and Refresh Learning	155
6.6	Parallel VLSI architecture for Fuzzy ART and VQ, including template learning and refresh functions.	156
6.7	Circuit schematic of the Fuzzy ART and VQ template matching cell, with integrated learning and template refresh. The dashed inset indicates a matched double pair of lateral bipolar transistors.	159
6.8	Layout of the Fuzzy ART and VQ template matching cell, of size 71 μm $\times$ 71 μm in 2 μm CMOS technology.	160

6.9	Centroid geometry of the matched double pair of lateral bipolar transistors, in conventional n-well CMOS technology.	161
6.10	Simplified schematic of the learning and refresh circuitry, in common for a column of Fuzzy ART/VQ cells. Analog multiplexers are implemented with complementary CMOS switches.	162
6.11	Chip micrograph of the 16×16 array, analog learning Fuzzy ART classifier and VQ. The die size is 2.2×2.25 mm^2 in 2 μm CMOS technology.	163
6.12	Measured distance output for one row of cells, sweeping one input component while fixing the other 15 inputs.	164
6.13	Stability of the analog memory array. Drift over 1000 self-refresh cycles, for 300 different initial conditions.	165
6.14	Template adaptation of Fuzzy ART in learning mode, under fixed inputs, from low and high template initial conditions.	166
7.1	Schematic Representation of Physical Part	170
7.2	Morphology of a QRS complex	173
7.3	Block Diagram for ART based Auto Alarm	175
C.1	Mismatch Characterization Chip Simplified Schematic	212
C.2	Experimentally measured/extracted mismatch data (diamonds) as a function of transistor size, for NMOS transistors. Also shown are the interpolated surfaces. (a) Results for $\sigma_{(\Delta\beta/\beta)}$, (b) for $\sigma_{(\Delta V_{TO})}$, (c) and for $\sigma_{(\Delta\gamma)}$	220

List of Tables

1.1	126 values of the 22 observable features of a mushroom database	9
1.2	Number of Training Iterations and Resulting Categories	34
3.1	Precision of the WTA	68
3.2	Delay Times of the WTA	70
3.3	Classification of Faults detected in the 30 fabricated chips	75
3.4	Description of Faults	76
3.5	Current Mismatch Components for Transistor Arrays with $10\mu A$ nominal current, $10\mu m \times 10\mu m$ Transistor Size, and $2.5mm \times 2.5mm$ die area, for the ES2-1.0μm CMOS process.	79
3.6	Measured Current Error for L_{A1} NMOS Current Sources	80
3.7	Measured Current Error for L_{A2} NMOS Current Sources	82
3.8	Measured Current Error for L_B PMOS Current Sources	82
4.1	Simulated Transient Times	101
4.2	Current-Mode WTA Precision Measurements	113
4.3	WTA Precision Computations (obtained through Hspice simulations) for the circuit reported in [Choi, 1993]	114
4.4	Measured delay times for a one chip WTA in ES2 technology	114
4.5	Measured delay times for a one chip WTA in MIETEC technology	114
4.6	Measured delay times for a two-chips WTA	115
7.1	Raman Scattered Wavelengths for Glucose with $\lambda_0 = 780nm$	170
C.1	Examples of Mismatch Models in the Literature ($x = 1/W, y = 1/L$)	211
C.2	NMOS Mismatch Characterization Parameters averaged over all measured dies and indicating the $\pm 3\sigma$ spread from die to die	217
C.3	PMOS Mismatch Characterization Parameters averaged over all measured dies and indicating the $\pm 3\sigma$ spread from die to die	218
C.4	NMOS Precision Measurements for the data in Table C.2	219
C.5	PMOS Precision Measurements for the data in Table C.3	221

C.6	NMOS Resulting coefficients for the fitting functions		222
C.7	PMOS Resulting coefficients for the fitting functions		222

Preface

Adaptive Resonance Theory (ART) is a well established neural network theory developed by Carpenter, Grossberg, and their collaborators at Boston University and the Center for Adaptive Systems. The development of the theory behind adaptive resonance began in 1976 with a work by Stephen Grossberg on the analysis of human cognitive information processing and stable coding in a complex input environment. ART neural network models have added a series of new principles to the original theory and have realized these principles as quantitative systems that can be applied to problems of category learning, recognition and prediction. The first ART neural network model appeared in the open literature in 1987 and is known as ART1. It is an unsupervised learning neural clustering architecture whose inputs are patterns composed of binary valued pixels and it groups them into categories according to a similarity criterion based on Hamming distances, modulated by a variable coarseness vigilance criterion. As a result, a set of extraordinary mathematical properties arise, rarely present in other algorithms of similar functionality. Almost simultaneously to the first ART1 publication, a similar algorithm named ART2 was published intended to cluster input patterns composed of analog valued pixels. It was relatively more complicated than ART1. But in 1991 a Fuzzy–ART architecture was reported which extended the original ART1 functionality by generalizing its operators using fuzzy set theory concepts. The result is that Fuzzy–ART can take patterns with analog-valued pixels as input while keeping the original mathematical properties present in ART1. Almost simultaneously, the ART1 and Fuzzy–ART architectures were extended to ARTMAP and Fuzzy–ARTMAP, respectively, which are supervised learning architectures that can be trained to learn the correspondence between an input pattern and the class to which it belongs, similar to what the popular Back-Propagation (BP) algorithm can do. The advantage of these ARTMAP architectures with respect to BP are mainly that they converge in a few training epochs (while BP needs of the order of thousands to even hundreds of thousands) and they are able to learn more complicated problems which BP cannot.

When the ART1 architecture was first reported, it was presented as a neural system described by a set of Short Term Memory (STM) and a set of Long Term Memory (LTM) nonlinear and coupled time domain differential equations. STM equations described the instantaneous activation evolution for the neurons as a function of the externally applied inputs and the present set of interconnection weights, while LTM equations described the time evolution of the adaptive interconnection weights which store the knowledge and experience of the complete system. STM equations settle much faster than LTM equations. This type of description can be called a *Type*-1 description. In an ART system, if the input patterns are held stable long enough so that STM equations reach their steady state, this steady state can be computed directly without solving the STM differential equations. The STM steady state can be obtained by solving a set of algebraic equations, properly sequenced. A *Type*-2 description of an ART system is one in which the STM state is computed by solving algebraic equations, and the LTM evolution is computed by solving the corresponding differential equations. If input patterns are held constant long enough so that both STM and LTM equations settle to their respective states, then the ART system operation can be described by solving properly sequenced algebraic equations only. Such a description can be called *Type*-3 and corresponds to the particular case called *Fast Learning* in the original ART1 paper. When the Fuzzy–ART, ARTMAP, and Fuzzy–ARTMAP algorithms were reported, they were described in their *Type*-3 or *Fast Learning* version, or, at the most, a *Slow Learning* LTM update was considered in which finite difference equations instead of differential equations are used. In this book, all ART systems are considered to be *Type*-3 or *Fast Learning* descriptions of the ART architectures.

In the open literature, all reported work on ART architectures and their applications is developed as software algorithms, running on conventional sequential computers. However, the parallel nature of these architectures and the simplicity of its components calls in a natural way for hardware implementations, similar to what nature has done with brains in living beings. Also, the fact that these ART, ARTMAP, and other architectures can be combined hierarchically to build higher level systems that solve complicated cognitive problems (as has been already done for example in robotics, vision systems and speech recognition), makes it even more attractive to develop a set of hardware components to be used in more complicated and hierarchically structured systems. There have been some attempts in the past to implement in hardware some of the aspects of ART architectures. However, they were intended to emulate *Type*-1 or *Type*-2 descriptions, and the results were bulky and inefficient pieces of hardware that could only realize part of the functionality of the powerful ART algorithms.

In 1992 Teresa Serrano–Gotarredona started her PhD at the National Microelectronics Center in Seville, Spain, supervised by Bernabé Linares-Barranco, who had finished his PhD the year before, and she would work on VLSI hardware implementations of ART1 and ARTMAP. In 1996 she presented an outstanding dissertation for which each chapter corresponds to at least one journal publication. The ART VLSI microchips developed during her PhD work implemented the full functionality of these systems, and the number of neurons and synapses was sufficiently high to demonstrate their potential application to real world problems and applications. The central journal publication for this work, "*A Real-Time Clustering Microchip Neural Engine*" published in the *IEEE Transactions on VLSI Systems* in June of 1996, received the 1995/96 Best Paper Award from the IEEE Circuits and Systems Society. As a consequence of this PhD work, Teresa and Bernabé, who got married at this moment, were offered to spend one year at the Johns Hopkins University in collaboration with professors Andreas G. Andreou and Gert Cauwenberghs who were engaged in a project with Stephen Grossberg. The objective was to further develop hardware realizations of ART and other systems produced by Stephen Grossberg's group at Boston University's Center for Adaptive Systems. The content of this book is mainly Teresa's PhD dissertation, condensed into three chapters and two appendixes, complemented with two chapters that describe some of the work Teresa and Bernabé did at Johns Hopkins University, and some other ART–oriented work that was being developed at Hopkins. Unfortunately, other ART–oriented work that was realized at Hopkins could not be included in this book due to editorial time constraints. In any case, besides the chapters corresponding to Teresa's dissertation, the other ART hardware information in this book describes work that is still going on at the moment of this writing, the results of which will hopefully be published in the near future. Besides these chapters, two more chapters have been added to this book for completeness, the first chapter on ART theory and the last chapter on applications. The book is structured as follows.

The first chapter is intended to introduce the unfamiliar reader to the *Fast Learning* versions of ART1, ARTMAP, Fuzzy–ART, and Fuzzy–ARTMAP architectures. This tutorial chapter is complemented with a very simple set of MATLAB programs that illustrate the operation of the four algorithms and can be used as well to easily test variants for these algorithms. The programs and their use are described in Appendix A.

Chapter 2 presents a variant of the original ART1 algorithm that makes it more VLSI–friendly. In essence, it consists of replacing a division operator with a subtraction operator. Appendix B demonstrates that all the mathematical properties of the original ART1 algorithm are preserved when performing this change. However, there are differences in behavior which are empirically studied in Chapter 2. Also, a mapping between parameters of the two algorithm versions is presented, which minimizes the difference in behavior between the two ART1 algorithms.

In Chapter 3, two VLSI microchips that implement the ART1 algorithm are presented. The first chip, developed first, uses a $1 cm^2$ chip fabricated in a $1.6 \mu m$ standard CMOS process and contains an ART1 system with 100 $F1$ nodes and 18 $F2$ nodes (which would be equivalent to 400×72 nodes in a scalable $0.35 \mu m$ CMOS process). After a transistor mismatch characterization of the technology, which is described in Appendix C, we were able to design a more efficient ART1 chip which was fabricated in a $1.0\mu m$ standard CMOS technology. The circuit technique used in this prototype, if expanded to a $1 cm^2$ chip, could host an ART1 system with 150 $F1$ nodes and 130 $F2$ nodes (equivalent to 375×325 nodes in a scalable $0.35\mu m$ CMOS process). In these chips, computations are performed with analog circuitry, although inputs and outputs as well as weight storage is done digitally. Therefore, it sounds reasonable to call these chips "*RAM Processors*", since they are basically a big RAM with special circuitry which performs some extra processing with the stored information and the externally applied patterns.

The chips in Chapter 3 have a small number of $F2$ layer nodes. For practical applications a relatively higher number of these nodes is desirable. The chips in Chapter 3 can be assembled in a modular fashion to increase the number of $F2$ nodes. However, the Winner–Takes–All $F2$ layer circuit would need to be distributed among several chips. The Winner-Takes-All circuits reported in the specialized literature at that time suffer from great degradation when distributed among several chips. Chapter 4 presents a new Winner–Takes–All circuit technique that does not suffer from this problem, even if the chips were fabricated by different foundries.

Chapter 5 describes a second generation *RAM Processor* ART chip. This design was started during the one year stay of Teresa and Bernabé at Hopkins, and is still going on at the time of this writing. It is a chip that can be configured as ART1, Fuzzy–ART, ARTMAP, and Fuzzy–ARTMAP. It has digital storage, and a $1cm^2$ microchip fabricated in a $0.5\mu m$ standard CMOS process could host an ART1 or ARTMAP system with 385 $F1$ nodes and 430 $F2$ nodes, or a Fuzzy–ART or Fuzzy–ARTMAP system with 77 $F1$ nodes and 430 $F2$ nodes (equivalent to 135×750 nodes in a scalable $0.35\mu m$ CMOS process).

In Chapter 6, professor Gert Cauwenberghs and collaborators describe some of the work they are undertaking on Fuzzy–ART systems. These systems differ from the others described in this book in that they do analog weight storage instead of digital. Technology inherent leakage currents do not permit non-volatile analog storage and this fact imposes the use of a weight refresh technique. The chips in this chapter implement variants of the theoretical Fuzzy–ART algorithm, in the sense that either the mathematical computations have been simplified to produce more efficient hardware, or the algorithm has been modified to cope with the problem of analog weight storage leakage. The resulting cell density would make it feasible to put a Fuzzy–ART system with 256 $F1$ nodes and 1024 $F2$ nodes in a $1cm^2$ chip fabricated in $0.35\mu m$ standard CMOS technology.

In the last chapter, three application examples from the literature have been identified, which illustrate examples for which the eventual use of ART microchips could be beneficial as compared to pure software implementations on conventional computers.

We hope that with this book people from industry and academia will find an easy path to become familiar with the powerful ART algorithms and discover a new dimension in the sense that it might be relatively simple to map these algorithms into highly efficient hardware with a high potential for application in autonomous intelligent systems where real time speed, limited power consumption and reduced sizes are constraints of primary concern. And, maybe in the future, we might witness a proliferation of ART system applications and the extensive use of *Adaptive Resonance Theory Microchips*.

To our respective sons Pablo,
Gonzalo and Gregory

1 ADAPTIVE RESONANCE THEORY ALGORITHMS

1.1 INTRODUCTION

Adaptive Resonance Theory is a well established neural network framework, developed at the Center for Adaptive Systems of Boston University. It is based on a solid study of mathematical models developed during many years [Grossberg, 1976], [Grossberg, 1980], [Carpenter, 1991a] and which made possible the invention of a series of architectures of Adaptive Resonance Theory (ART). There is an extensive variety of ART architectures. Furthermore, new architectures are being reported as we write these lines, and certainly more will appear in the future.

In this book we will concentrate on hardware implementations of only four of these ART architectures, namely ART1, ARTMAP, Fuzzy–ART, and Fuzzy–ARTMAP. The main characteristics of each of them are:

- ART1 [Carpenter, 1987]: unsupervised learning clustering architecture for binary-valued input patterns.

- ARTMAP [Carpenter, 1991b]: supervised learning classification architecture for binary-valued input pairs of patterns (or pattern-class pairs).

- Fuzzy-ART [Carpenter, 1991c]: unsupervised learning clustering architecture for analog-valued input patterns.

2   ADAPTIVE RESONANCE THEORY MICROCHIPS

- Fuzzy-ARTMAP [Carpenter, 1992]: supervised learning classification architecture for analog-valued input pairs of patterns (or pattern-class pairs).

Historically, the architectures for binary valued input patterns were developed first and later extended to operate on analog valued input patterns. The latter ones reduce exactly to the former ones if binary valued input patterns are provided. Thus it would not make much sense to study them separately since the former ones are actually a particular case of the latter ones. However, from a hardware implementation point of view it makes a big difference to build a clustering engine for binary or for analog valued input patterns. In the present book hardware VLSI systems will be described for both type of architectures. Consequently, in this Chapter we will briefly describe each of them separately.

Although in some of the original ART papers the architectures are described using a time domain nonlinear and coupled system of differential equations, in this book we will concentrate only on their algorithmic description. It is this type of descriptions that have been mapped into hardware in later Chapters. To illustrate and better understand these algorithms some "home made" MATLAB code examples have been included in Appendix A.

1.2   ART1

The ART1 architecture is a massively parallel neural network pattern recognition machine which self organizes recognition codes in response to a sequence of binary valued input patterns [Carpenter, 1987]. The system receives a sequence of binary valued input patterns and will cluster them into a set of categories in an unsupervised way.

Topological Description of the ART1 Architecture

Fig. 1.1 shows the topological structure of an ART1 system. It consists of two layers of neurons or processing cells named *F1* and *F2*. Each neuron in layer *F1* receives the value of an input pattern pixel. If there are N neurons in layer *F1*, the input pattern can have up to N pixels I_i ($i = 1, ..., N$). Pixels I_i are binary valued. For convenience let us use the values '0' and '1'. Input patterns presented to layer *F1* are going to be clustered into categories, and a neuron in layer *F2* will represent a possible category. Each neuron in layer *F1* is connected to all neurons in layer *F2* through bottom-up synaptic connections of strength z_{ij}^{bu}. Index i indicates that the connection goes from the *i-th* neuron in layer *F1* to the *j-th* neuron in layer *F2*. Bottom-up weights z_{ij}^{bu} are of analog nature and they may take any value within the bounded interval [0,1]. The input to the *j-th F2* neuron is given by

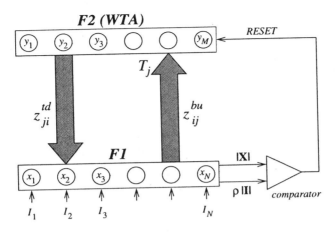

Figure 1.1. Topological structure of the ART1 architecture

$$T_j = \sum_{i=1}^{N} z_{ij}^{bu} I_i \quad , \quad j = 1, \ldots, M \tag{1.1}$$

where we are assuming there are M $F2$ nodes. Terms T_j are called *Choice Functions* and represent a certain "distance" between the stored pattern $\mathbf{z}_j^{bu} \equiv (z_{1j}^{bu}, z_{2j}^{bu}, \ldots, z_{Nj}^{bu})$ and input pattern $\mathbf{I} \equiv (I_1, I_2, \ldots, I_N)$.

Neurons in layer $F2$ operate in such a way that their output y_j is always '0', except for the neuron receiving the largest T_j input from the $F1$ layer. This $F2$ neuron, let us call it J, has output '1',

$$\begin{aligned} y_J &= 1 \quad if \quad T_J = max_j\{T_j\} \\ y_{j \neq J} &= 0 \end{aligned} \tag{1.2}$$

Each $F2$ neuron is connected to all $F1$ neurons through top-down synaptic connections of strength z_{ji}^{td}, which are binary-valued (i.e., z_{ji}^{td} is either '0' or '1'). Thus the i-th $F1$ neuron input from the $F2$ layer is

$$V_i = \sum_{j=1}^{M} z_{ji}^{td} y_j = z_{Ji}^{td} \quad , \quad i = 1, \ldots, N \tag{1.3}$$

There is a vigilance subsystem, formed by the comparator in Fig. 1.1, which checks the appropriateness of the active $F2$ category. This subsystem compares the norm of vector $\mathbf{X} \equiv (x_1, \ldots, x_N)$, defined as

$$x_i = V_i I_i \quad or \quad \mathbf{X} = \mathbf{V} \cap \mathbf{I} = \mathbf{z}_J^{td} \cap \mathbf{I} \qquad (1.4)$$

to the norm of vector $\rho \mathbf{I}$, where $\rho \in [0, 1]$ is the vigilance parameter. Depending on the result of this comparison the vigilance subsystem may reset the actual active $F2$ category. The norm of a vector $\mathbf{a} \equiv (a_1, \ldots, a_N)$ is the l_1 norm

$$|\mathbf{a}| = \sum_{i=1}^{N} |a_i| \qquad (1.5)$$

ART1 Dynamics

The time evolution of the state of all $F1$ and $F2$ neurons is governed by a set of time domain nonlinear and coupled differential equations, called Short Term Memory equations, and the present state of $F1$ and $F2$ neurons is called Short Term Memory (STM). The time domain evolution of the set of weights z_{ij}^{bu} and z_{ji}^{td} is governed by another set of time domain nonlinear differential equations called Long Term Memory equations, and the set of values stored in weights z_{ij}^{bu} and z_{ji}^{td} is called Long Term Memory (LTM). The time constant associated to the LTM equations is much slower than that of the STM equations. Consequently, if an input pattern $\mathbf{I}$ is presented to the $F1$ layer, STM will settle first. If the input pattern $\mathbf{I}$ is held constant at the $F1$ layer input until all STM equations and the vigilance subsystem settle, it is possible to describe the STM dynamics using an algebraic description of the steady state of the STM differential equations. Furthermore, if the input pattern $\mathbf{I}$ is held constant until LTM settles, then it is also possible to describe the LTM dynamics using algebraic equations that define the steady state of the LTM differential equations. In this case the dynamic description of the ART1 architecture is referred to as "*Fast Learning* ART1". From here on we will concentrate on this type of description. The *Fast Learning* algorithmic description of the ART1 architecture is shown in Fig. 1.2. Note that only two parameters are needed, ρ which is called the *Vigilance Parameter* and takes a value between '0' and '1', and parameter L which takes a value larger than '1'.

First, all interconnection weights are initialized. These weights store the knowledge or experience of the ART1 system. Therefore, after initialization they do not hold any information on categories, clusters, nor past input patterns provided. Bottom-up weights are initialized to $z_{ij}^{bu} = L/(L - 1 + N)$ and top-down weights to $z_{ji}^{td} = 1$. Now the system is ready to receive its first external input pattern $\mathbf{I} = (I_1, \ldots, I_N)$ where I_i may be either '0' or '1'. At this point, the input to each $F2$ neuron can be computed,

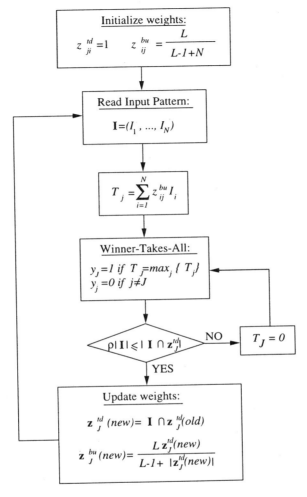

Figure 1.2. Algorithmic description of ART1 functionality

$$T_j = \sum_{i=1}^{N} z_{ij}^{bu} I_i \quad , \quad j = 1, \ldots, M \tag{1.6}$$

In the next step the $F2$ neuron receiving the largest input T_j is activated, while all others are deactivated. Thus, if T_J is the maximum of all T_j inputs, then $y_J = 1$ and $y_{j \neq J} = 0$. Once an $F2$ node is active, the vigilance subsystem checks if it is appropriate. The vigilance subsystem action is characterized by the vigilance parameter ρ which is set to a value between '0' and '1'. If the following condition is satisfied

$$\rho \leq \frac{|\mathbf{I} \cap \mathbf{z}_J^{td}|}{|\mathbf{I}|} \tag{1.7}$$

where

$$|\mathbf{I}| = \sum_{i=1}^{N} I_i \quad , \quad |\mathbf{I} \cap \mathbf{z}_J^{td}| = \sum_{i=1}^{N} I_i z_{Ji}^{td} \tag{1.8}$$

the active $F2$ category node J is selected for LTM update. Otherwise, $F2$ node J is shut off (by forcing $T_J = 0$) and a new $F2$ node will become active. This node is again checked by the vigilance subsystem and shut off if it does not satisfy the condition of eq. (1.7). And so on until an active $F2$ node meets this vigilance criterion. Once this node is found the bottom-up and top-down connection weights related to this node are updated

$$\begin{aligned} z_{Ji}^{td}(new) &= z_{Ji}^{td}(old) I_i \\ z_{iJ}^{bu}(new) &= \frac{L z_{Ji}^{td}(new)}{L - 1 + |\mathbf{z}_J^{td}(new)|} \end{aligned} \tag{1.9}$$

or, in vector notation

$$\begin{aligned} \mathbf{z}_J^{td}(new) &= \mathbf{z}_J^{td}(old) \cap \mathbf{I} \\ \mathbf{z}_J^{bu}(new) &= \frac{L \mathbf{z}_J^{td}(new)}{L - 1 + |\mathbf{z}_J^{td}(new)|} \end{aligned} \tag{1.10}$$

Now the system is ready to receive the next input pattern.

If an $F2$ category node j has not yet been chosen for category storage it is said to be an *uncommitted* node and its weights z_{ij}^{bu} and z_{ji}^{td}, $i = (1, ..., N)$, still preserve their initialized values. On the other hand, if an $F2$ node has already been selected, at least once, for storage, it is referred to as a *committed* node. Note that initially, since all weights z_{ij}^{bu} are equal, the first time all $F2$ inputs T_j are computed with equation eq. (1.6), they will be identical and it would not be possible to choose a maximum among them. This can be solved by making $M = n_c + 1$, where n_c is the number of committed nodes. This way, initially $n_c = 0$ and $M = 1$, which means that only the first $F2$ node is available for competition. As soon as this node is chosen for storage $n_c = 1$ and $M = 2$, so that next time the competition will be between one committed and one uncommitted node. Once this second node is chosen for storage $n_c = 2$ and $M = 3$. And so on. The competition in the $F2$ layer will always be between the n_c committed nodes and one uncommitted node. Note also that an uncommitted node always satisfies the vigilance criterion of eq. (1.7) for any $\rho \in [0, 1]$, because

$$\frac{|\mathbf{I} \cap \mathbf{z}_{j_{uncommitted}}^{td}|}{|\mathbf{I}|} = \frac{|\mathbf{I}|}{|\mathbf{I}|} = 1 \geq \rho \tag{1.11}$$

Therefore if, at a given time, the maximum T_J corresponds to an uncommitted node, this node will be chosen for storage and become committed.

As an illustrative example, Fig. 1.3 shows how the top-down $\mathbf{z}_j^{td}$ templates change as the sequence of input patterns (depicted in the left side of the figure) are presented to the system. By eq. (1.9) the bottom-up weights are a normalized version of the top-down weights and, therefore, do not contain any additional information. This is why in Fig. 1.3 we only show the top-down weights because they fully characterize their corresponding $F2$ category node. At the bottom of Fig. 1.3, below each final category $\mathbf{z}_j^{td}$, are written the input patterns that belong to each of the final categories. For example, input patterns 1 and 7 have been clustered into category $\mathbf{z}_1^{td}$, and so on. Appendix A shows a MATLAB code that implements this example.

In Fig. 1.3 each input pattern $\mathbf{I}$ is drawn as a square matrix as if it were some kind of bitmap image. Input patterns of an ART1 system can indeed be bitmap images, but can also be more general. For example, Table 1.1 shows 22 observable features of a mushroom, according to a benchmark learning database [Schlimmer, 1987] used in [Carpenter, 1991b]. Each feature can have several characteristics or possible values. For example, feature number 1 'cap-shape' can be 'bell', 'conical', 'convex', 'flat', 'knobled', or 'sunken'. Therefore, this feature can be characterized by 6 bits. If the first bit is '1' it means cap-shape is 'bell'. If the second bit is '1', cap-shape is 'conical', and so on. This way mushrooms can be clustered into different categories according to the 22 observable features of Table 1.1. Each feature would be represented by a number of bits (6 bits for 'cap-shape', 4 bits for 'cap-surface', ...), resulting in 126 bits of which only 22 would be '1' and the rest '0'. These 126-bit feature vectors can be used as inputs $\mathbf{I}$ to an ART1 system to produce clusters for Mushrooms categorizations.

Some interesting ART1 properties

The ART1 architecture possesses some interesting properties which compares it favorably against other clustering algorithms. Some of these properties are:

8 ADAPTIVE RESONANCE THEORY MICROCHIPS

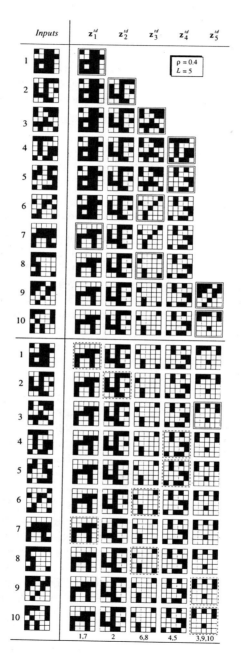

Figure 1.3. Training sequence of ART1 with randomly generated input patterns

Table 1.1. 126 values of the 22 observable features of a mushroom database

Number	Feature	Possible Values
1	Cap-Shape	Bell, Conical, Convex, Flat, Knobbed, Sunken
2	Cap-Surface	Fibrous, Grooves, Scaly, Smooth
3	Cap-Color	Brown, Buff, Gray, Pink, Purple, Red, White, Yellow, Cinnamon
4	Bruises	Bruises, No Bruises
5	Odor	None, Almond, Anise, Creosote, Fishy, Foul, Musty, Pungent, Spicy
6	Gill-Attachment	Attached, Descending, Free, Notched
7	Gill-Spacing	Close, Crowded, Distant
8	Gill-Size	Brode, Narrow
9	Gill-Color	Brown, Buff, Orange, Gray, Green, Pink, Purple, Red, White, Yellow, Chocolate, Black
10	Stalk-Shape	Enlarging, Tapering
11	Stalk-Root	Bulbous, Club, Cup, Equal, Rhizomorphs, Rooted, Missing
12	Stalk-Surface-Above-Ring	Fibrous, Silky, Scaly, Smooth
13	Stalk-Surface-Below-Ring	Fibrous, Silky, Scaly, Smooth
14	Stalk-Color-Above-Ring	Brown, Buff, Orange, Gray, Pink, Red, White, Yellow, Cinnamon
15	Stalk-Color-Below-Ring	Brown, Buff, Orange, Gray, Pink, Red, White, Yellow, Cinnamon
16	Veil-Type	Partial, Universal
17	Veil-Color	Brown, Orange, White, Yellow
18	Ring-Number	None, One, Two
19	Ring-Type	None, Cobwebby, Evanescent, Flaring, Large, Pendant, Sheathing, Zone
20	Spore-Print-Color	Brown, Buff, Orange, Green, Purple, White, Yellow, Chocolate, Black
21	Population	Abundant, Clustered, Numerous, Scattered, Several, Solitary
22	Habitat	Grasses, Leaves, Meadows, Paths, Urban, Waste, Woods

10 ADAPTIVE RESONANCE THEORY MICROCHIPS

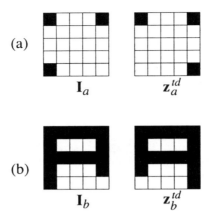

Figure 1.4. Illustration of the self-scaling property. (a) One mismatch out of three features, (b) one mismatch out of fifteen features

Vigilance or Variable Coarseness. One of the most characteristic features of the ART1 algorithm is the possibility to externally tune the coarseness with which categories should be formed. The ART1 architecture contains a *Vigilance Subsystem*, controlled by a *Vigilance Parameter* ρ, that provides this functionality.

The vigilance parameter ρ can be set to any value between '0' and '1'. The smaller ρ is set, the more input patterns will be clustered together into the same category, which means a higher generalization capability. The higher ρ is set, more attention will be paid to differences between input patterns and a larger number of categories will result.

By looking at the algorithmic description of ART1 in Fig. 1.2, one may wonder what the purpose of the vigilance subsystem might be. Why do we need to check whether or not

$$\rho|\mathbf{I}| \leq |\mathbf{I} \cap \mathbf{z}_J^{td}| \qquad (1.12)$$

if category J has already been selected by the Winner-Takes-All? The answer is related to the *Self-Scaling* property discussed next.

Self-Scaling. This property allows features to be treated as either noise or signal, according to context. For example, consider the two patterns $\mathbf{I}_a$ or $\mathbf{z}_a^{td}$ shown in Fig. 1.4(a). They differ only in one feature, but it is one out of three. Therefore, it seems reasonable to classify the two patterns in Fig. 1.4(a) as belonging to two different categories. In this case, the feature that makes the difference is said to be a 'critical feature' or signal. For the case of Fig. 1.4(b) the two patterns $\mathbf{I}_b$ and $\mathbf{z}_b^{td}$ differ also in one feature, but now it is one mismatch out of 15. Therefore, it seems reasonable to classify them into the same category. In this case, the feature that makes the difference is said to be 'noise'.

If in Fig. 1.4 the patterns on the left side were new input patterns, $\mathbf{I}_a$ and $\mathbf{I}_b$, and the patterns on the right side were the ones that characterize some stored category, $\mathbf{z}_a^{td}$ and $\mathbf{z}_b^{td}$, let us see how the algorithm of Fig. 1.2 would proceed in each case.

For the case of Fig. 1.4(a) if pattern $\mathbf{I}_a$ is presented to the $F1$ layer and the category characterized by weight template $\mathbf{z}_a^{td}$ received the largest input T_J, then the vigilance subsystem would accept this category for pattern $\mathbf{I}_a$ if

$$\rho \leq \frac{|\mathbf{I}_a \cap \mathbf{z}_a^{td}|}{|\mathbf{I}_a|} = \frac{2}{3} = 0.6666 \tag{1.13}$$

Consequently, if parameter ρ were initially set to a value higher than 0.666, pattern $\mathbf{I}_a$ would not be stored into the category characterized by weight template $\mathbf{z}_a^{td}$.

For the case of Fig. 1.4(b) the condition would change to

$$\rho \leq \frac{|\mathbf{I}_b \cap \mathbf{z}_b^{td}|}{|\mathbf{I}_b|} = \frac{14}{15} = 0.9333 \tag{1.14}$$

So, if for example the vigilance parameter were $\rho = 0.85$ patterns $\mathbf{I}_b$ and $\mathbf{z}_b^{td}$ would have been clustered together, but patterns $\mathbf{I}_a$ and $\mathbf{z}_a^{td}$ not. These examples also illustrate the role of vigilance parameter ρ. The larger or closer it is to '1', the more categories will be formed and more attention will be paid to the 'details' that distinguish the input patterns. The smaller ρ is, or closer to '0', less attention will be paid to 'little details' and more input patterns will be clustered into the same categories, resulting in a smaller number of categories and more generalization capability, for the same sequence of input patterns.

Self-Stabilization in a Small Number of Iterations. All the interconnection weights in the system that are subject to learning reach a stationary value after a finite number of presentations of a sequence of arbitrarily many and arbitrarily complex binary input patterns.

Initially, when a category j is uncommitted, it stores a template with all its elements $z_{ji}^{td} = 1$. When category j becomes committed by an input pattern $\mathbf{I}$, it looses the elements z_{ji}^{td} such that the corresponding $I_i = 0$,

$$z_{ji}^{td}(new) = z_{ji}^{td}(old) I_i \tag{1.15}$$

Hence, each category j progressively looses elements. However, the number of elements a category can loose is at most N. Consequently, if we denote as n_j the number of times the weight template $\mathbf{z}_j^{td}$ of category j is submitted to change, then

$$n_j \leq N \tag{1.16}$$

If there are M categories in the system, the number of times the weights in the system are changed is limited to a maximum of $N \times M$.

In practical examples, the system weights always stabilize in a reduced number of input pattern presentations.

On-line Learning. For many clustering algorithms, given a set of exemplars or input patterns, the clusters or categories are computed *Off-Line*. If a new exemplar needs to be added, then the system knowledge has to be erased and retrained with the updated database. *Off-Line* learning means that *learning phase* and *performing phase* are separate phases. In ART1 this does not happen. ART1 can be trained *On-Line*, this is, it learns while it performs. Every time a new input pattern is given ART1 will answer with a category (either committed or uncommitted) and update this category to incorporate the new knowledge. This on-line learning property makes the ART1 an ideal candidate for real-time clustering.

Capturing Rare Events. The ART1 algorithm is able to learn and form clusters with input exemplars that appear very rarely. Thanks to its *On-Line* learning capability, ART1 can learn a rare input pattern with only a single exemplar presentation. Since the pattern is rare it will solicit an uncommitted node. The rest of patterns, since they differ significantly from this rare one, will never choose its category for update, and consequently will not alter it.

Direct Access to Familiar Input Patterns. An input pattern **I** is said to have *Direct Access* to a stored category j, if this category is the first one chosen by the *F2* (WTA) layer and it meets the vigilance criterion.

As the human cognitive system, ART1 has the ability to quickly recognize an object which is familiar to the system. No matter how many recognition codes (or categories) the system may store, after stabilization the system always accesses directly the code of patterns that have been previously learned or which are very similar to other previously learned input patterns.

Direct Access to Subset and Superset Patterns. Suppose that a learning process has produced a set of categories in the *F2* layer. Suppose that two of these categories, j_1 and j_2, are such that $\mathbf{z}_{j_1}^{td} \subset \mathbf{z}_{j_2}^{td}$ (this means that if $z_{j_1 i}^{td} = 1$ then it must be $z_{j_2 i}^{td} = 1$, but if $z_{j_1 i}^{td} = 0$ then $z_{j_2 i}^{td}$ can either be '0' or

'1'). In this case we say that $\mathbf{z}_{j_1}^{td}$ is a *Subset Template* of $\mathbf{z}_{j_2}^{td}$, or equivalently, $\mathbf{z}_{j_2}^{td}$ is a *Superset Template* of $\mathbf{z}_{j_1}^{td}$. Mathematically, in vector notation, this can be expressed as,

$$\mathbf{z}_{j_1}^{td} \cap \mathbf{z}_{j_2}^{td} = \mathbf{z}_{j_1}^{td} \qquad (1.17)$$

Consider two input patterns $\mathbf{I}^{(1)}$ and $\mathbf{I}^{(2)}$ such that,

$$\begin{aligned} \mathbf{I}^{(1)} &= \mathbf{z}_{j_1}^{td} \equiv (z_{j_1 1}^{td}, z_{j_1 2}^{td}, \ldots, z_{j_1 N}^{td}) \\ \mathbf{I}^{(2)} &= \mathbf{z}_{j_2}^{td} \equiv (z_{j_2 1}^{td}, z_{j_2 2}^{td}, \ldots, z_{j_2 N}^{td}) \end{aligned} \qquad (1.18)$$

The *Direct Access to Subset and Superset* property assures that input $\mathbf{I}^{(1)}$ has *Direct Access* to category j_1 and that input $\mathbf{I}^{(2)}$ has *Direct Access* to category j_2.

First, suppose input $\mathbf{I}^{(1)}$ is presented to the system. Let us compute the values of T_{j_1} and T_{j_2},

$$T_{j_1} = \frac{L \sum_{i=1}^{N} I_i^{(1)} z_{j_1 i}^{td}}{L - 1 + |\mathbf{z}_{j_1}^{td}|} = \frac{L |\mathbf{I}^{(1)} \cap \mathbf{I}^{(1)}|}{L - 1 + |\mathbf{I}^{(1)}|} = \frac{L |\mathbf{I}^{(1)}|}{L - 1 + |\mathbf{I}^{(1)}|} \qquad (1.19)$$

$$T_{j_2} = \frac{L \sum_{i=1}^{N} I_i^{(1)} z_{j_2 i}^{td}}{L - 1 + |\mathbf{z}_{j_2}|} = \frac{L |\mathbf{I}^{(1)} \cap \mathbf{I}^{(2)}|}{L - 1 + |\mathbf{I}^{(2)}|} = \frac{L |\mathbf{I}^{(1)}|}{L - 1 + |\mathbf{I}^{(2)}|} \qquad (1.20)$$

Since $|\mathbf{I}^{(1)}| < |\mathbf{I}^{(2)}|$, it is obvious that $T_{j_1} > T_{j_2}$ (remember $L > 1$) and therefore category j_1 will be selected by the $F2$ layer. This category will also be accepted by the vigilance subsystem because

$$\frac{|\mathbf{I}^{(1)} \cap \mathbf{z}_{j_1}^{td}|}{|\mathbf{z}_{j_1}^{td}|} = \frac{|\mathbf{I}^{(1)}|}{|\mathbf{I}^{(1)}|} = 1 \geq \rho \qquad (1.21)$$

On the other hand, if input pattern $\mathbf{I}^{(2)}$ is presented at the input,

$$T_{j_1} = \frac{L \sum_{i=1}^{N} I_i^{(2)} z_{j_1 i}^{td}}{L - 1 + |\mathbf{z}_{j_1}^{td}|} = \frac{L |\mathbf{I}^{(2)} \cap \mathbf{I}^{(1)}|}{L - 1 + |\mathbf{I}^{(1)}|} = \frac{L |\mathbf{I}^{(1)}|}{L - 1 + |\mathbf{I}^{(1)}|} \qquad (1.22)$$

$$T_{j_2} = \frac{L \sum_{i=1}^{N} I_i^{(2)} z_{j_2 i}^{td}}{L - 1 + |\mathbf{z}_{j_2}^{td}|} = \frac{L |\mathbf{I}^{(2)} \cap \mathbf{I}^{(2)}|}{L - 1 + |\mathbf{I}^{(2)}|} = \frac{L |\mathbf{I}^{(2)}|}{L - 1 + |\mathbf{I}^{(2)}|} \qquad (1.23)$$

Since function $Lx/(L-1+x)$ is an increasing function with x, it results that $T_{j_2} > T_{j_1}$ and category j_2 will be chosen by the $F2$ layer, and accepted by the vigilance subsystem since

$$\frac{|\mathbf{I}^{(2)} \cap \mathbf{z}_{j_2}^{td}|}{|\mathbf{z}_{j_2}^{td}|} = \frac{|\mathbf{I}^{(2)}|}{|\mathbf{I}^{(2)}|} = 1 \geq \rho \qquad (1.24)$$

Biasing the Network to Form New Categories. Independently of the vigilance parameter ρ, parameter 'L' biases the tendency of the network to form a smaller or larger number of categories. In particular, parameter L biases the tendency of the network to select a new uncommitted category before a committed one. When an input pattern **I** is presented, an uncommitted node is chosen before a committed one j if

$$\frac{|\mathbf{I} \cap \mathbf{z}_j^{td}|}{L-1+|\mathbf{z}_j^{td}|} < \frac{|\mathbf{I}|}{L-1+N} \qquad (1.25)$$

This inequality is equivalent to

$$L - 1 > \frac{N|\mathbf{I} \cap \mathbf{z}_j^{td}| - |\mathbf{z}_j^{td}||\mathbf{I}|}{|\mathbf{I}| - |\mathbf{I} \cap \mathbf{z}_j^{td}|} \qquad (1.26)$$

Therefore, increasing L increases the tendency to select an uncommitted node before a committed one.

1.3 ARTMAP

ARTMAP is an Adaptive Resonance Theory based architecture that learns, in a supervised way, associations of pairs of patterns [Carpenter, 1991b]. Two possible realizations can be implemented, which we will call here *Simple ARTMAP* and *General ARTMAP*.

Simple ARTMAP

This ARTMAP implementation has a structure which consists of an ART1 module and a *Map-Field* module as shown in Fig. 1.5. Binary input patterns $\mathbf{a} = (a_1, a_2, \ldots, a_{N_a})$ are fed to the $F1$ layer of the ART1 module, while the class vector $\mathbf{y}^b = (y_1^b, y_2^b, \ldots, y_{M_b}^b)$ is such that all its components are '0' except one which is set to '1'. Let us call K the class which is set to '1' for a given input pattern $\mathbf{a}$,

ADAPTIVE RESONANCE THEORY ALGORITHMS   15

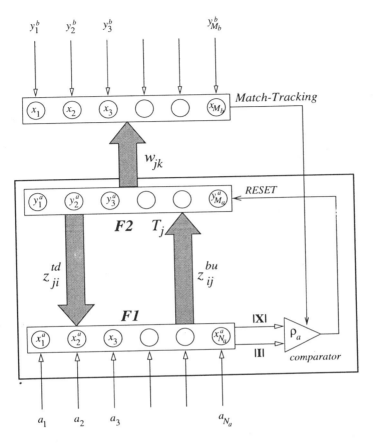

Figure 1.5.   Simple ARTMAP architecture

$$y_K = 1 \quad (1.27)$$
$$y_{k \neq K} = 0$$

where $k = 1, \ldots, M_b$. This means that the ARTMAP system should learn that pattern **a** belongs to class K.

For example, it could be the case that we want to teach ARTMAP to learn, for each letter of the alphabet, many possible exemplars like upper and lower case, italic or bold representations, different fonts, and so on. After training, ARTMAP should be able to tell us which letter we are giving it, independently of its representation and even allowing some degree of distortion or added noise. This means that ARTMAP has *generalization* capability. Fig. 1.6 shows, for letters 'A', 'B', and 'C', different 10 × 10 pixel possible representations. Each representation is a 100-bit **a** input pattern. Its associated $\mathbf{y}^b$ class vector would

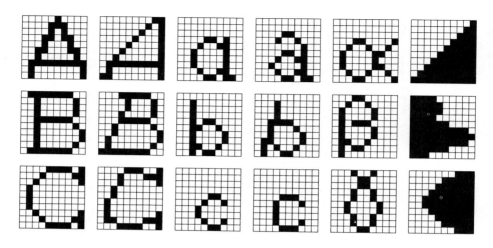

Figure 1.6. Different exemplars for letters 'A', 'B', and 'C'

be $(1, 0, 0)$ for a letter 'A' character, $(0, 1, 0)$ for a letter 'B' character, or $(0, 0, 1)$ for a letter 'C' character.

The ART1 module will try to establish compressed codes for the characters in Fig. 1.6 as long as they belong to the same class or letter representation. The *Map-Field* module is responsible for mapping the resulting ART1 module categories y^a to the externally provided set of classes y^b. For example, if for the six letter 'A' representations of Fig. 1.6 the ART1 module is able to generate three clusters, the *Map-Field* module has to map these three clusters to $\mathbf{y}^b$ pattern $(1, 0, 0)$, which represents letter 'A'.

Learning Operation Mode. As shown in Fig. 1.5, the simplest ARTMAP system consists of one ART1 module with an N_a nodes $F1$ layer and an M_a nodes $F2$ layer, a *Map-Field* module with M_b nodes, a matrix of $M_a \times M_b$ map-field weights w_{jk}, and a *Match-Tracking* subsystem that controls the ART1 module vigilance parameter ρ_a. The way this ARTMAP system operates can be observed in the flow diagram of Fig. 1.7.

First the ART1 weights z_{ij}^{bu}, z_{ji}^{td} and map-field weights w_{jk} are initialized. The next step is to provide an initial value for the ART1 module vigilance parameter ρ_a. Let us call this initial value $\overline{\rho_a}$. Vigilance parameter ρ_a may increase above this value, but never will be less than this initial value. Next an input pattern $\mathbf{a}$ is read as well as the class it belongs to, defined by the *pattern-class* pair $(\mathbf{a}, \mathbf{y}^b)$. The following steps are pure ART1 operations, and identify an ART1 winning category y_J^a that meets the present ρ_a vigilance criterion. Once this ART1 category J is identified the *Match-Tracking* subsystem starts to work. The given class vector $\mathbf{y}^b$ is such that component K is '1' ($y_K^b = 1$)

ADAPTIVE RESONANCE THEORY ALGORITHMS

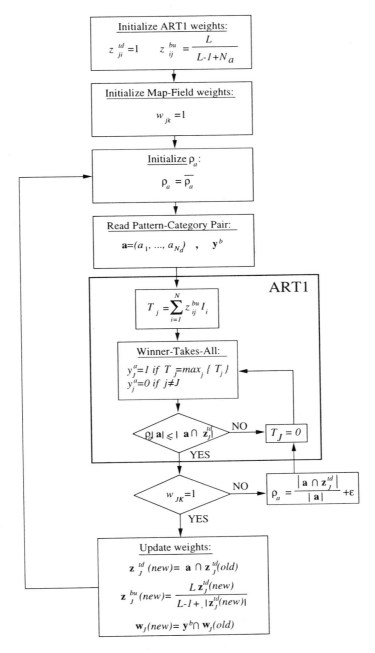

Figure 1.7. Flow diagram for simple ARTMAP architecture learning operation mode

and all others are '0' ($y^b_{k \neq K} = 0$). If map-field weight w_{JK} is '1' it means that ART1 cluster J satisfies one of the two following conditions:

1. it has been associated previously to class K, or

2. it has not been associated previously to any class K (because the ART1 cluster has just been formed).

However, if map-field weight w_{JK} is '0' it means that ART1 cluster J has been previously associated to a class different than K. This implies that the actual input pattern **a** should not be clustered into category J, because this category is already associated to a class different than K. In this case the *Match-Tracking* subsystem increases the ART1 module vigilance parameter ρ_a until category J is shut off. A ρ_a value that fulfills this condition is

$$\rho_a = \frac{|\mathbf{a} \cap \mathbf{z}^{td}_J|}{|\mathbf{a}|} + \epsilon \quad (1.28)$$

where $\epsilon > 0$ but close to zero. Now another ART1 category will be selected and the *Match-Tracking* subsystem will again verify if it is appropriate. If not ρ_a will be increased again, and so on until a valid ART1 cluster is identified, or a new ART1 cluster is formed. Note that if a new ART1 cluster is formed it will be accepted by the *Match-Tracking* subsystem because in this case $w_{Jk} = 1$, $\forall$ k.

Once a valid ART1 category is available, weights will be updated for both the ART1 module and the *Map-Field* module

$$\mathbf{z}^{bu}_J(new) = \frac{L(\mathbf{a} \cap \mathbf{z}^{td}_J(old))}{L - 1 + |\mathbf{a} \cap \mathbf{z}^{td}_J(old)|}$$

$$\mathbf{z}^{td}_J(new) = \mathbf{a} \cap \mathbf{z}^{td}_J(old) \quad (1.29)$$

$$\mathbf{w}_J(new) = \mathbf{y}^b \cap \mathbf{w}_J(old)$$

Recall Operation Mode. If only input pattern **a** is given to an ARTMAP system but no class vector $\mathbf{y}^b$ is supplied, then ARTMAP estimates a class for pattern **a**, based on its previous experience and generalization capability. In this case the algorithm of Fig. 1.7 is simplified to the one shown in Fig. 1.8. Note that now there is no *Match-Tracking* and no weights update (and, of course, no weights initialization). The ART1 module vigilance parameter ρ_a is set to the initial value $\overline{\rho_a}$ used during training. Under these circumstances the ARTMAP output is class vector $\mathbf{y}^b$ which is made equal to vector $\mathbf{w}_J$,

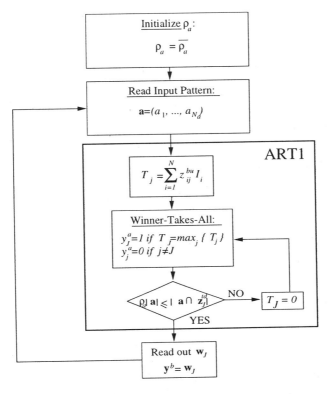

Figure 1.8. Flow diagram for simple ARTMAP architecture recall operation mode

$$y_k^b = w_{Jk} \quad , \quad k = 1, \ldots, M_b \tag{1.30}$$

There are two possible cases:

1. Vector $\mathbf{y}^b$ has only one component set to '1', while all others a '0',

$$\begin{aligned} y_K^b &= 1 \\ y_{k \neq K}^b &= 0 \end{aligned} \tag{1.31}$$

This means that the actual input pattern $\mathbf{a}$ has been recalled as belonging to class K.

2. Vector $\mathbf{y}^b$ has all of its components set to '1',

$$y_k^b = 1 \quad , \quad \forall k \tag{1.32}$$

This means that the actual input pattern **a** wants to establish a new ART1 category, because for vigilance parameter $\overline{\rho_a}$ none of the available ART1 categories are satisfactory. This corresponds to a *"Don't know"* answer, because pattern **a** differs too much from any prior experience.

On-Line Learning versus Off-Line Learning. An ARTMAP system can either operate with *On-Line* learning or *Off-Line* learning.

In *Off-Line* learning a fixed training set is repeatedly presented to the system until all weights stabilize. When this occurs the ARTMAP system is able to recall the proper class vector $\mathbf{y}^b$ for each pattern **a** of the training set. After the training stage learning (weights update) is shut off and only patterns **a** are provided to the system, so that ARTMAP will give as output a class vector $\mathbf{y}^b$ for each **a** pattern presentation. The answer can either be one of the possible classes used during training, or a *"Don't know"* answer if pattern **a** is not sufficiently similar to any one used in the training set.

Another way the ARTMAP system can be made to operate is with *On-Line* learning. In this mode training and recall operations can occur at any time. There is no fixed training set, and the system can either be exposed to an *"Input Pattern"*-*"Class Vector"* pair $(\mathbf{a}, \mathbf{y}^b)$, or simply to an isolated *"Input Pattern"*. In the former case ARTMAP will learn the association according to the diagram of Fig. 1.7, while in the latter case ARTMAP will provide a possible class vector $\mathbf{y}^b$ for the isolated *"Input Pattern"* according to the diagram of Fig. 1.8. In this mode the system knowledge will increase with experience and more precise answers will be given as ARTMAP is exposed to more $(\mathbf{a}, \mathbf{y}^b)$ pairs. This operation mode is similar to human beings operating in natural environments.

Complement Coding Option. Sometimes it is useful that the norm of all input patterns **a** be the same. For ART based systems with analog inputs (like Fuzzy-ART and Fuzzy-ARTMAP) this avoids the category proliferation problem described by Moore [Moore, 1989] and solved by Carpenter and Grossberg using complement coding [Carpenter, 1991c], [Carpenter, 1992]. In ARTMAP, where **a** is of binary nature, another problem might appear if the algorithm of Fig. 1.7 is applied to input patterns of different norms. For example, in the case of Fig. 1.6 the third font for letter 'C' (let us call it $\mathbf{I}_{C3}$) is a subset template of the last font of letter 'A' (let us call it $\mathbf{I}_{A6}$). If $\overline{\rho_a}$ is sufficiently high then pattern $\mathbf{I}_{A6}$ is learned and kept untouched until $\mathbf{I}_{C3}$ is given. Since $\mathbf{I}_{C3} \cap \mathbf{I}_{A6} = \mathbf{I}_{C3}$, pattern $\mathbf{I}_{C3}$ will want to cluster with $\mathbf{I}_{A6}$. However, they are exemplars of different classes and the match-tracking system will increase ρ_a. By eq. (1.28) the new value for the vigilance parameter should be $\rho_a = 1 + \epsilon > 0$ and no input pattern would satisfy this vigilance criterion: the algorithm of Fig. 1.8 would not work[1].

This problem does not appear if $|\mathbf{a}|$ is constant for all input patterns. Therefore, if $|\mathbf{a}|$ is not constant, complement coding is necessary in ARTMAP to avoid this problem.

Complement coding consists in duplicating the number of components of the input pattern. Instead of using $\mathbf{a} \equiv (a_1, a_2, \ldots, a_{N_a})$ as input pattern, the $2N_a$-dimensional binary vector

$$(\mathbf{a}, \mathbf{a}^c) \equiv (a_1, a_2, \ldots a_{N_a}, a_1^c, a_2^c, \ldots, a_{N_a}^c) \qquad (1.33)$$

is used, where

$$a_i^c = 1 - a_i \qquad (1.34)$$

Sometimes, however, it is not necessary to use complement encoding because the norm of all input patterns $|\mathbf{a}|$ is already constant or approximately similar. For example, in the case described earlier in Table 1.1, where a 126 components vector describes 22 observable features of a mushroom, the norm of each input vector is always 22. In that example complement coding is superfluous.

In Appendix A a MATLAB code example for ARTMAP is included. The code implements the example presented in Fig. 1.6 and uses complement coding.

General ARTMAP Architecture

As opposed to the simple ARTMAP architecture described in the previous Section and composed of one ART1 module, a *Map-Field* module and a *Match-Tracking* subsystem, there is a more general ARTMAP version which consists of two ART1 modules, the *Map-Field* module, and the *Match-Tracking* subsystem, as shown in Fig. 1.9. With this architecture pairs of input patterns (**a**,**b**) are given to the system. Each ART1 subsystem will establish clusters for both type of input patterns **a** and **b**, and the *Map-Field* subsystem will learn the corresponding associations between those clusters. For example, such an ARTMAP could be trained to learn the associations of visual representation of fruits, vegetables, ... and their corresponding tastes. Each ART1 subsystem will try to achieve a certain code compression in its memory. The more code compression is achieved by each ART1 subsystem, the greater is the system generalization capability.

---

[1] A 'dirty' solution to this would be to check if $\rho_a > 1$ after applying eq. (1.28). If it is true, then skip to step '*initialize* $\overline{\rho_a}$' without any update. The algorithm would continue working but will not be able to learn the exemplar.

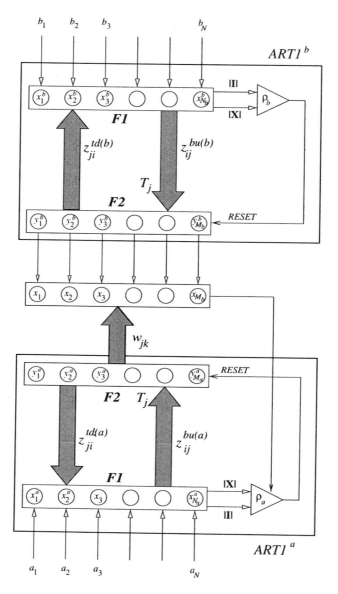

Figure 1.9. General ARTMAP architecture

Some ARTMAP Interesting Properties

Since ARTMAP includes one or two ART1 modules it inherits all ART1 properties. But due to its *Map-Field* structure and *Match-Tracking* system other properties become available, some of which will be highlighted next.

1. As in ART1, ARTMAP is able to learn rare events, but it is able to do it even if they are very similar to other more frequent events as long as their consequences are different. If two very similar (but distinct) patterns have to be learned as belonging to separate classes, the *match-tracking* system will increase ρ_a until this is achieved. And this is done whether or not one or both of these input patterns are rare or frequent.

2. When using *Fast Learning*, learning usually stabilizes in a few training trials. Thus ARTMAP learns to make accurate predictions quickly. Even if learning has not completely stabilized, ARTMAP is able to perform acceptable predictions. This allows the use of this architecture with "*On-Line*" learning.

3. Learning is stable. This allows to update its internal knowledge continuously, and permits to learn one or more new databases without eroding prior knowledge.

4. Memory capacity can be chosen arbitrarily large without sacrificing the stability of fast learning or its accurate generalization capability.

5. The ARTMAP *match-tracking* system has the ability to conjointly maximize generalization and minimize prediction error on a trial-by-trial basis, using only local operations.

1.4 FUZZY-ART

Fuzzy-ART is a clustering neural network architecture which self-organizes recognition codes in response to sequences of analog or binary input patterns.

Fuzzy-ART Architecture

The Fuzzy-ART architecture is shown in Fig. 1.10. It has the same structure than the ART1 system shown in Fig. 1.1. It consists of two layers of computing cells or neurons *F1* and *F2*, and a vigilance subsystem controlled by an adjustable vigilance parameter $\rho \in [0, 1]$.

Layer *F1* is the input layer composed of N input cells. Each input cell receives a component $I_i \in [0, 1]$ of the analog input vector $\mathbf{I} = (I_1, \ldots, I_N)$. Layer *F2* is the category layer. It is composed of M cells, each one representing a possible category. Each category cell receives an input T_j. Each *F1* layer neuron i is connected to each *F2* layer neuron j by a synaptic connection of weight z_{ij}^{bu}. Each *F2* layer neuron j is connected to each *F1* layer neuron i by a synaptic connection of strength z_{ji}^{td}. In Fuzzy-ART $z_{ij}^{bu} = z_{ji}^{td}$. Consequently, from now on we will refer to the weights as $z_{ij} = z_{ij}^{bu} = z_{ji}^{td}$.

The main differences between the ART1 and Fuzzy-ART architectures are:

- The input vectors are analog valued. That is $\mathbf{I} = (I_1, \ldots, I_N)$ is an N-dimensional vector with each component $I_i \in [0, 1]$.

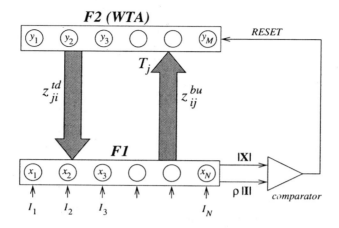

Figure 1.10. Topological structure of the Fuzzy-ART architecture

- There is only one set of analog valued weight vectors $\mathbf{z}_j = (z_{1j}, \ldots, z_{Nj})$ $j = 1, \ldots, M$.

- In the computation of the choice functions T_j, the learning rule, and the vigilance criterion: the intersection operation $\cap$ (binary AND) is substituted by the fuzzy MIN operator $\wedge$ (analog AND).

Fuzzy-ART Operation

Figure 1.11 shows the flow diagram of the Fuzzy-ART operation. Initially, all the interconnection weights z_{ij} are set to '1'.

When an analog input vector $\mathbf{I} = (I_1, \ldots, I_N)$ is applied to the system, each *F1* layer cell receives a component $I_i \in [0, 1]$. Then each *F2* layer category cell receives an input T_j, which is a measurement of the similarity between the analog input pattern $\mathbf{I}$ and the analog weight template $\mathbf{z}_j = (z_{1j}, \ldots, z_{Nj})$ stored in category j,

$$T_j = \frac{|\mathbf{I} \wedge \mathbf{z}_j|}{\alpha + |\mathbf{z}_j|} \tag{1.35}$$

where $\wedge$ is the fuzzy MIN operator defined by $(\mathbf{X} \wedge \mathbf{Y})_i = \min(X_i, Y_i)$, $|\mathbf{X}|$ is the l_1 norm $|\mathbf{X}| = \sum_{i=1}^{N} |X_i|$, and α is a positive parameter called 'choice parameter'.

The *j-th F2* cell gives an output y_j which is '1' if this cell is receiving the largest T_j input and '0' otherwise.

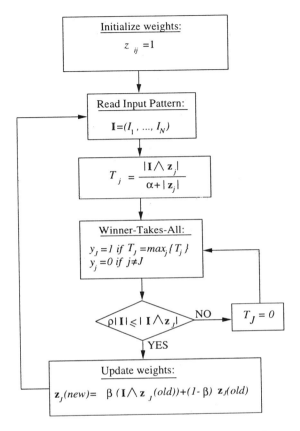

Figure 1.11. Flow Diagram of the Fuzzy-ART Operation

$$y_J = 1 \quad \text{if} \quad T_J = \max\{T_j\}$$
$$y_{j \neq J} = 0 \quad \text{otherwise} \tag{1.36}$$

This way, the *F2* layer (which acts as a Winner-Takes-All) selects the category J whose stored template $\mathbf{z}_J$ most closely resembles input pattern $\mathbf{I}$ according to the similarity criterion defined by eq. (1.35).

For the winning category J, the vigilance subsystem checks the condition,

$$\rho|\mathbf{I}| \leq |\mathbf{I} \wedge \mathbf{z}_J| \tag{1.37}$$

If this condition is not true, category J is discarded by forcing $T_J = 0$. Layer *F2* will again select the category with maximum T_j, and the vigilance criterion of eq. (1.37) will be checked again. This search process continues until layer $F2$ finds a winning category which fulfills the vigilance criterion.

When a category J that meets the vigilance criterion is activated, its weights $\mathbf{z}_J$ are updated according to the rule

$$\mathbf{z}_J(new) = \beta(\mathbf{I} \wedge \mathbf{z}_J(old)) + (1-\beta)\mathbf{z}_J(old) \qquad (1.38)$$

where β is the 'learning rate parameter' $\beta \in [0,1]$.

Fast-Commit Slow-Recode Option

For efficient coding of noisy input sets, it is useful to set $\beta = 1$ when the learning category J is an uncommitted node (fast-commit) and $\beta < 1$ after the category is committed (slow-recode).

With this option, the first time category J becomes active $\mathbf{z}_J(new) = \mathbf{I}$, allowing an adequate response to inputs that may occur only rarely and in response to which a quick and accurate performance may be needed.

When a committed category needs to be updated $\beta < 1$, thus preventing features that have been incorporated into it from being deleted when a noisy or partial input appears. Only a persistent change in a feature allows to delete it from a category template.

Input Normalization Option

A category proliferation problem may occur in some ART systems when the norm of input vectors $|\mathbf{I}|$ can be made arbitrarily small [Moore, 1989], [Carpenter, 1991c]. This problem of category proliferation is avoided in the Fuzzy-ART system when the input vectors are normalized before being processed by the system [Carpenter, 1991c].

An input vector $|\mathbf{I}|$ is said to be normalized when there exists a constant $\gamma > 0$, such that $|\mathbf{I}| = \gamma$ for all input vectors $\mathbf{I}$. One way of normalizing the input vectors could be to divide each incoming vector $\mathbf{a}$ by its norm $\mathbf{I} = \mathbf{a}/|\mathbf{a}|$. However, this method may loose the information about the input amplitude. Consider, for example, the two-dimensional incoming vectors $\mathbf{a}_1 = (1,1)$ and $\mathbf{a}_2 = (0.1, 0.1)$. The first vector $\mathbf{a}_1$ indicates a high value of the two vector components, while the second vector $\mathbf{a}_2$ indicates a low value of both vector components. However, both incoming vectors $\mathbf{a}_1$ and $\mathbf{a}_2$ will produce the same normalized input vector $\mathbf{I} = (1/2, 1/2)$ and will be treated by the system in the same way.

To avoid this loss of information, Grossberg et al.[Carpenter, 1991c] proposed the complement coding rule for the normalization of the input vectors. This rule consists of expanding an N-dimensional incoming vector $\mathbf{a} = (a_1, \ldots, a_N)$ to a $2N$-dimensional vector defined by

$$\mathbf{I} = (\mathbf{a}, \mathbf{a}^c) = (a_1, \ldots, a_N, a_1^c, \ldots, a_N^c) \qquad (1.39)$$

where $a_i^c = 1 - a_i$ for $i = 1, \ldots, N$. This way, all the input vectors $\mathbf{I}$ are normalized

$$|\mathbf{I}| = |(\mathbf{a}, \mathbf{a}^c)| = \sum_{i=1}^{N} a_i + \sum_{i=1}^{N} a_i^c = \sum_{i=1}^{N} a_i + N - \sum_{i=1}^{N} a_i = N \qquad (1.40)$$

but the amplitude information is preserved. The two incoming vectors $\mathbf{a}_1$ and $\mathbf{a}_2$ discussed above will produce two different input vectors $\mathbf{I}_1 = (1, 1, 0, 0)$ and $\mathbf{I}_2 = (0.1, 0.1, 0.9, 0.9)$ and will be treated by the system in a different way.

In the case of a Fuzzy-ART system with the complement coding option, the weight vectors $\mathbf{z}_j$ are also expanded to $2N$-dimensional vectors,

$$\mathbf{z}_j = (\mathbf{u}_j, \mathbf{v}_j^c) \quad j = 1, \ldots, M \qquad (1.41)$$

which are initially set to $\mathbf{z}_j = (1, \ldots, 1)$, so that $\mathbf{u}_j = (1, \ldots, 1)$ and $\mathbf{v}_j = (0, \ldots, 0)$. The same learning rule given by eq. (1.38) is still valid for updating the $\mathbf{z}_j$ vectors.

Geometric Interpretation of the Fuzzy-ART Learning with Complement Coding

Let the input patterns be two-dimensional vectors $\mathbf{a} = (a_1, a_2)$. By complement coding, the effective input vectors are four-dimensional $\mathbf{I} = (a_1, a_2, 1 - a_1, 1 - a_2)$. In this case, each category j will be represented by a four-dimensional weight vector

$$\mathbf{z}_j = (\mathbf{u}_j, \mathbf{v}_j^c) \qquad (1.42)$$

where $\mathbf{u}_j$ and $\mathbf{v}_j$ are two-dimensional vectors. Consider the rectangle R_j with corners defined by vectors $\mathbf{u}_j$ and $\mathbf{v}_j$. The size of the rectangle R_j can be defined as

$$|R_j| = |\mathbf{v}_j - \mathbf{u}_j| \qquad (1.43)$$

Let us assume the system is in the fast learning mode, that is, $\beta = 1$ in eq. (1.38). When a category becomes committed for the first time by an input pattern $\mathbf{I} = (\mathbf{a}, \mathbf{a}^c)$, that category learns the template

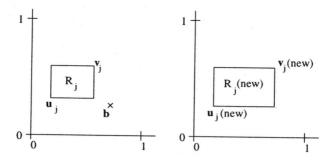

Figure 1.12. Expansion of the rectangle R_j when a new input vector **b** is incorporated into category j

$$\mathbf{z}_j(new) = (\mathbf{a}, \mathbf{a}^c) \tag{1.44}$$

so that $\mathbf{u}_j = \mathbf{v}_j = \mathbf{a}$ and the rectangle R_j is just point **a** and it has zero size.

When a new input pattern $\mathbf{I} = (\mathbf{b}, \mathbf{b}^c)$ is added to a category j, rectangle R_j is expanded (according to the learning rule given by eq. (1.38) with $\beta = 1$) to

$$\begin{aligned} \mathbf{z}_j(new) &= (\mathbf{u}_j(new), \mathbf{v}_j^c(new)) = \\ &= (\mathbf{u}_j(old) \wedge \mathbf{b}, \mathbf{v}_j^c(old) \wedge \mathbf{b}^c) = \\ &= (\mathbf{u}_j(old) \wedge \mathbf{b}, (1 - \mathbf{v}_j(old)) \wedge (1 - \mathbf{b})) = \\ &= (\mathbf{u}_j(old) \wedge \mathbf{b}, (\mathbf{v}_j(old) \vee \mathbf{b})^c) \end{aligned} \tag{1.45}$$

where symbol $\vee$ denotes the component wise MAX operator. Therefore,

$$\begin{aligned} \mathbf{u}_j(new) &= \mathbf{u}_j(old) \wedge \mathbf{b} \\ \mathbf{v}_j(new) &= \mathbf{v}_j(old) \vee \mathbf{b} \end{aligned} \tag{1.46}$$

Fig. 1.12 illustrates how rectangle R_j is expanded when category j incorporates a new input vector. As shown in Fig. 1.12, rectangle R_j is expanded by the minimum size needed to incorporate the new input vector **b** inside the rectangle. In particular, if **b** is an input vector inside R_j no weight change occurs during the weight update.

The maximum size that a rectangle R_j can reach is limited by the vigilance parameter ρ. This can be reasoned as follows. If an input vector $\mathbf{I} = (\mathbf{b}, \mathbf{b}^c)$ activates a category j, this category will be reset whenever,

$$|\mathbf{I} \wedge \mathbf{z}_j| < \rho |\mathbf{I}| \tag{1.47}$$

Since input vectors are two dimensional and complement coding is used, $|\mathbf{I}| = N = 2$. Hence, the reset condition becomes,

$$|\mathbf{I} \wedge \mathbf{z}_j| < 2\rho \tag{1.48}$$

but,

$$\begin{aligned}
|\mathbf{I} \wedge \mathbf{z}_j| &= |(\mathbf{b}, \mathbf{b}^c) \wedge (\mathbf{u}_j, \mathbf{v}_j^c)| = \\
&= |\mathbf{b} \wedge \mathbf{u}_j| + |(\mathbf{b} \vee \mathbf{v}_j)^c| = \\
&= |\mathbf{b} \wedge \mathbf{u}_j| + 2 - |\mathbf{b} \vee \mathbf{v}_j| = \\
&= |\mathbf{u}_j(new)| + 2 - |\mathbf{v}_j(new)| = 2 - |R_j(new)|
\end{aligned} \tag{1.49}$$

Therefore, the category will be reset whenever,

$$|R_j^{new}| > 2(1 - \rho) \tag{1.50}$$

and the maximum size of the rectangles will be limited by $2(1 - \rho)$ in the 2-dimensional case. For input vectors with N components ($2N$ after complement coding) the maximum size rectangle is limited by $N(1 - \rho)$. Consequently, the closer ρ is to '1' the smaller the size of the rectangles R_j will be and the smaller the number of input patterns coded in each category.

The fact that rectangles grow during learning and that their maximum size is bounded allows the existence of a *Stable Category Learning Theorem* [Carpenter, 1991c] which guarantees that no category proliferation will occur. If no complement coding is used, the input space *"rectangle covering"* may be substituted by a *"triangle covering"* [Carpenter, 1991c]. But it turns out that the resulting triangles have a size which depends directly on the norm of their weight templates $|\mathbf{z}_J|$. This means that as $\mathbf{z}_J$ shrinks its associate triangle will shrink as well, and thus triangles close to the origin will be small. Consequently, the number of triangles needed to *"cover"* the input space close to the origin grows. This together with the fact that triangles may shrink during learning produces the category proliferation problem if input patterns are not normalized [Carpenter, 1991c].

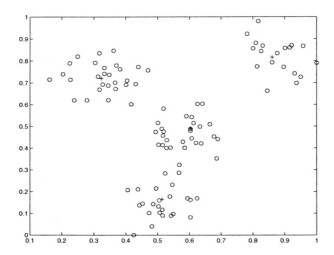

Figure 1.13. Set of Points used for Training of Fuzzy-ART

Example of Fuzzy-ART Categorization. Fig. 1.13 shows a set of $\mathcal{R}^2$ data points (circles) generated randomly around four central points (crosses). Fuzzy-ART was trained to cluster these points using different vigilance parameter ρ settings, while maintaining $\alpha = 0.1$. The results are shown in Fig. 1.14. For $\rho = 0.3$ all data points are clustered into two categories, shown in Fig. 1.14(a) with circles and dots, respectively. Also shown are the category boxes $(\mathbf{u}, \mathbf{v}^c)$. For $\rho = 0.6$ four categories result, as is shown in Fig. 1.14(b). Fig. 1.14(c) shows the five resulting categories for $\rho = 0.7$, and Fig. 1.14(d) shows the 10 resulting categories for $\rho = 0.8$. These figures were generated using the Fuzzy-ART MATLAB code example included in Appendix A.

1.5 FUZZY-ARTMAP

The Fuzzy-ARTMAP architecture is identical to the ARTMAP one, except that the ART1 module (or modules) have been substituted by Fuzzy-ART structures. Consequently, the only difference is that input patterns $\mathbf{a}$ (for the first Fuzzy-ART module) or $\mathbf{b}$ (for the optional second Fuzzy-ART module) are of analog nature. Each Fuzzy-ART module will generate compressed recognition codes (or categories) for the sequence of analog input patterns it receives. These recognition codes or categories are represented by category vectors $\mathbf{y}^a$ and $\mathbf{y}^b$ which have all of their components equal to '0', except for one which is equal to '1'. This is the same than for the ARTMAP architecture, and therefore,

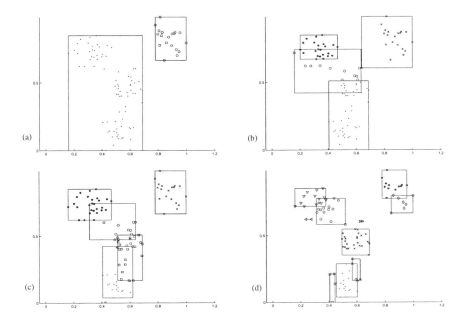

Figure 1.14. Categories formed by a Fuzzy-ART system trained with the training set shown in Fig. 1.13 with parameters $\alpha = 0.1$ and (a) $\rho = 0.3$, (b) $\rho = 0.6$, (c) $\rho = 0.7$, and (d) $\rho = 0.8$.

the Map-Field module behaves in the same manner for ARTMAP than for Fuzzy-ARTMAP.

As in ARTMAP one can distinguish between a simple Fuzzy-ARTMAP architecture, with only one Fuzzy-ART module, and a general Fuzzy-ARTMAP architecture, with two Fuzzy-ART modules. The only difference between them, is that for the general Fuzzy-ARTMAP the system will try to find compressed representations for the exemplars **b** given to the second Fuzzy-ART module. In the case of the simple Fuzzy-ART, the desired classes $\mathbf{y}^b$ are directly provided to the Map-Field module. This is the usual case, and from now on we will concentrate on this case unless stated otherwise.

Example: The Two Spiral Problem

The algorithmic description of the Fuzzy-ARTMAP architecture is shown if Fig. 1.15. As can be seen, it is identical to the ARTMAP flow diagram, except that the intersection operator is substituted by the minimum operator.

Since Fuzzy-ART is an analog Adaptive Resonance Theory architecture, in order to avoid the category proliferation problem [Moore, 1989], it is convenient

32 ADAPTIVE RESONANCE THEORY MICROCHIPS

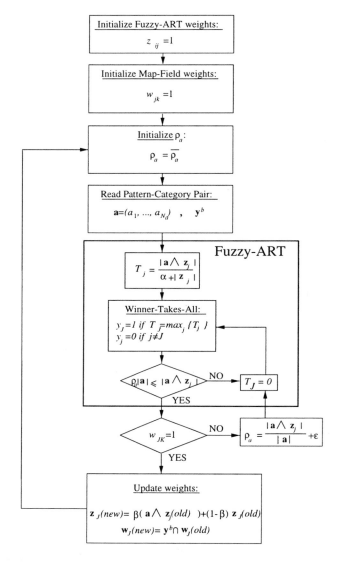

Figure 1.15. Flow Diagram for Fuzzy-ARTMAP Architecture

to normalize the input patterns **a**, in the same way as discussed for the Fuzzy-ART architecture. This can be done by either dividing all components by the norm of the input vector $\mathbf{a}/|\mathbf{a}|$ or by using complement coding

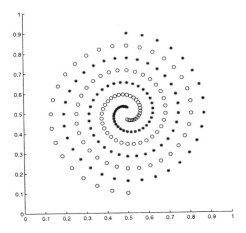

Figure 1.16. The two spirals used to train Fuzzy-ARTMAP

$$(\mathbf{a}, \mathbf{a}^c) \equiv (a_1, a_2, \ldots, a_N, a_1^c, a_2^c, \ldots, a_N^c)$$

$$a_i^c = 1 - a_i$$

(1.51)

Complement coding requires more system resources (twice more *F1* Fuzzy-ART nodes and weights) but treats the inputs in a more general case. As an illustrative example of the Fuzzy-ARTMAP operation let us see how it performs on a difficult benchmark problem: learning to tell two spirals apart. Consider the two spirals in $\mathcal{R}^2$ shown in Fig. 1.16. The $\mathcal{R}^2$ plane is considered to be divided into two regions. The first spiral are sample point of the first region, while the second spiral provides sample point for the second region. This benchmark problem, [Carpenter, 1992] tests the system's learning and generalization ability.

Each spiral makes three complete turns, with 32 points per turn plus an endpoint, totaling 97. During one training epoch, the outermost point of the first spiral is given first, then the outermost point of the second spiral, and so on, working in to the center of each spiral. If $(\mathbf{a}^l, \mathbf{b}^l)$, $l = 1, 2, \ldots, 2 \times 97$ with $\mathbf{a}^l = (a_1^l, a_2^l) \in \mathcal{R}^2$ and $\mathbf{b}^l = (b_1^l, b_2^l) \in \mathcal{R}^2$, is the sequence of exemplars, then for $n = 1, 2, \ldots, 97$

$$a_1^{2n-1} = 1 - a_1^{2n} = r_n \sin(\alpha_n) + 0.5$$

$$a_2^{2n-1} = 1 - a_2^{2n} = r_n \cos(\alpha_n) + 0.5$$

(1.52)

where

Table 1.2. Number of Training Iterations and Resulting Categories

$\overline{\rho_a}$	Iterations	Categories
0.000	5	30
0.400	5	30
0.500	4	30
0.600	3	33
0.700	3	38
0.800	3	39
0.850	2	40
0.900	1	45
0.950	1	78
1.000	1	194

$$\begin{aligned} r_n &= 0.4 \tfrac{105-n}{104} \\ \alpha_n &= \tfrac{\pi(n-1)}{16} \\ \mathbf{b}^{2n-1} &= (1,0) \ \text{(first spiral)} \\ \mathbf{b}^{2n} &= (0,1) \ \text{(second spiral)} \end{aligned} \quad (1.53)$$

In this example the system is trained with $\alpha = 0.5$ and for different values of $\overline{\rho_a}$ until all exemplars are perfectly learned. After that a 100×100 grid in the $\mathcal{R}^2$ plane is tested to see how Fuzzy-ART generalizes from the 2×97 exemplar points. Fig. 1.17 shows the results obtained from 9 cases where $\overline{\rho_a}$ took the values $\overline{\rho_a} = \{1.0, 7/8, 6/8, \ldots, 1/8, 0.0\}$ and $\alpha = 0.001$. For $\overline{\rho_a} < 0.4$ the results are identical to the case $\overline{\rho_a} = 0.4$. Appendix A includes the MATLAB code used to run this example.

Table 1.2 shows the number of training epochs needed for each case, as well as the resulting number of category nodes in the Fuzzy-ART module. As can be seen, an optimum code compression is achieved with $\overline{\rho_a} = 0.85$, while for lower values of $\overline{\rho_a}$ the performance is degraded. For $\overline{\rho_a}$ below 0.85 what is happening is that the training exemplars tend to create new categories in each training epoch which they tend to abandon in the next epoch (because it became too degraded after clustering together too many exemplars). The result is that a high number of categories appear, most of which are abandoned categories, i.e. they are not accessed by any of the training exemplars.

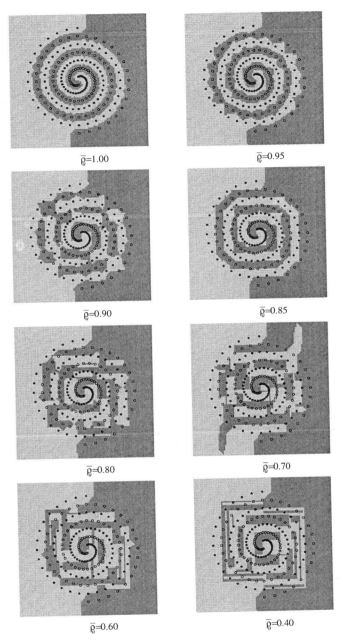

Figure 1.17. Test set results of Fuzzy-ARTMAP for the two spiral problem for different $\overline{\rho_a}$ values and $\alpha = 0.001$. The training set are the spiral points and the test set is the 100×100 grid points

Comparison to Back-Propagation

Lang and Witbrock [Lang, 1989] tried to train a Back Propagation network to learn to tell these two spirals apart. They were unable to do it using a conventional Back-Propagation system with connections from one layer to the next. However, they accomplished the task by using a 2-5-5-5-1 system such that each node was connected to all nodes in all subsequent layers, resulting in a system with 138 trainable weights. Weights were updated using either Vanilla Back-Propagation or the Quickprop algorithm [Fahlman, 1989]. On average, Vanilla Back-Propagation required 20000 epochs to learn to distinguish all points in the training set, while Quickprop required 8000 epochs. For Fuzzy-ARTMAP, when $\overline{\rho_a} = 0$ and $\alpha = 0.001$, it learns to solve the problem in 5 epochs using 30 Fuzzy-ART categories[2] (150 weights).

Recognition Improvement

When using Fast Learning it is observed that in general different orderings of the training set yield different adaptive weights and recognition categories, even when the overall predictive accuracy is similar. The different category structures cause the set of test set items where errors occur to vary from one simulation to the next.

There are two ways the final learned weights set and categories can be made independent of the order of training patterns presentations, namely slow-learning and voting strategy.

Slow-Learning. Slow-Learning in Fuzzy-ARTMAP not only means that the Fuzzy-ART module(s) should work with slow learning but also the update of the Map-Field weights. In this case, the Fuzzy-ARTMAP algorithmic flow diagram of Fig. 1.15 is changed to the one shown in Fig. 1.18. Now, both parameters β and λ may be set to values between '0' and '1'.

Note that now a new parameter, the *map field vigilance parameter* $0 < \rho_{ab} \leq 1$ appears. When using slow-learning a larger number of input pattern sequence presentations are needed. The smaller β and λ the larger the number of training epochs. For each training epoch the ordering of the sequence of input patterns is randomly altered. Thus the resulting set of learned weights and categories should be independent of training sequence orderings.

Voting Strategy. An alternative method that also increases prediction accuracy and yields results independent of training sequence orderings is the voting

---

[2] In [Carpenter, 1992] it is stated that with $\overline{\rho_a} = 0$ and $\alpha \approx 0$ the resulting number of categories is 25. We have found (in MATLAB) that when $\alpha = 3 \times 10^{-15}$ then 25 categories result. However, around this value the resulting number of categories is very sensitive to α.

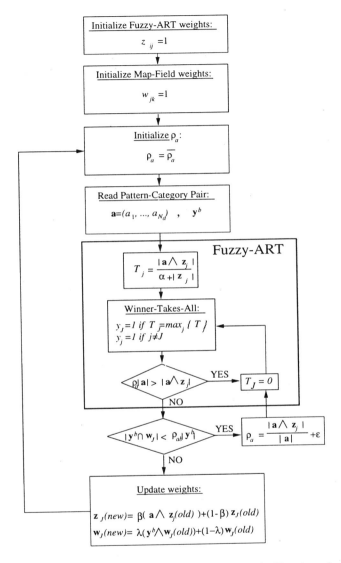

Figure 1.18. Flow Diagram of Fuzzy-ARTMAP Algorithm in Slow-Learning Mode

strategy based method. For this method several Fuzzy-ARTMAP systems are trained using Fast-Learning, but in such a way that the training sequence ordering is randomly altered for each system. When it comes to predict responses, each Fuzzy-ARTMAP system will do its own response. The final result is that predicted by the majority of Fuzzy-ARTMAP systems. Since the set of items making erroneous predictions varies from one system to another, voting cancels many of the errors. Furthermore, the voting strategy can be used to assign confidence estimates to competing predictions given by small, noisy, or incomplete training sets.

Fuzzy-ARTMAP Interesting Properties

This architecture presents interesting properties which are also found for the Fuzzy-ART or ARTMAP architectures. To highlight just a few, let us consider the following

1. Learning is stable. This property is achieved thanks to the fact that all weights (for both the Fuzzy-ART module z_{ij}^a, and the map field module w_{jk}) can only decrease in time. This is true whether Fast or Slow learning is used.

2. A small number of parameters are needed. These parameters are α and ρ_a if Fast learning is used, and for Slow learning β, λ, and ρ_{ab} need to be added. Furthermore, the system will always work independently of the values given to these parameters (provided they are within their specified ranges). Choosing different values for these parameters only produces different behavior (but always correct behavior).

3. No problem specific system crafting or initial conditions need to be provided. The architecture is always the same and the initial conditions are always identical, independently of the problem to be learned.

4. The match tracking subsystem which controls the value of vigilance parameter ρ_a, is such that it creates a minimal number of Fuzzy-ART^a categories needed to meet the accuracy criteria (controlled by ρ_a). It realizes a minimax learning rule that enables to learn quickly, efficiently, and accurately as it conjointly minimizes predictive error and maximizes predictive generalization. It automatically links predictive success to category size on a trial-by-trial basis using only local operations. It does this by increasing ρ_a by the minimal amount needed to correct the predictive error by the map field module.

5. As in Fuzzy-ART, there is a relation between the ρ_a value and the size of the established categories. The smaller $\overline{\rho_a}$ the more code compression is achieved and the more generalization capability is accomplished. On the other hand, the larger $\overline{\rho_a}$ is made, each Fuzzy-ART module category will have smaller size and smaller predictive error is expected, although a larger number of category nodes (and weights) will result. For $\overline{\rho_a} = 1$ each input pattern **a** will establish its own Fuzzy-ART category, which will make Fuzzy-ARTMAP to work as a nearest neighbor classifier.

6. As in ARTMAP, Fuzzy-ARTMAP can be trained either *Off-Line* (as in the examples discussed in this Section), or *On-Line*.

2 A VLSI-FRIENDLY ART1 ALGORITHM

When a specific computational algorithm needs to be implemented in hardware engineers have usually two options. First, identify the operations that are required and look for appropriate circuit operators that realize them or, second, spend a little extra time trying to simplify or reduce the operations required and implement a simplified theoretical model that hopefully preserves the original functionality.

This second one is the approach we used when we considered the hardware implementation possibility of the ART1 algorithm for the first time. This implementation is discussed in Chapter 3 and is based on the algorithm modification worked out in this Chapter and in Appendix B. The modification is basically the replacement of a division operation by a subtraction operation for computing the choice functions T_j. In this Chapter we introduce the reasonings that resulted in this simplification and we describe some statistical studies that compare the system performances with and without this simplification. Appendix B is a collection of theorems that prove the conservation of all ART1 mathematical properties after introducing the simplification. Although this Chapter concentrates exclusively on ART1, the simplification is extensible to the Fuzzy-ART architecture. Actually, for Fuzzy-ART (which is a generalization of ART1) several alternative choice functions have been introduced [Carpenter, 1994], one of them being the one discussed in this Chapter.

Later, in Chapter 5, a general ART chip design capable of realizing ART1, Fuzzy ART, ARTMAP, and Fuzzy ARTMAP is described where both choice function options (division-based or subtraction-based) are available. That chip however consumes significantly more circuit resources and design efforts because of this.

2.1 THE MODIFIED ART1 ALGORITHM

From a hardware implementation point of view, one of the first issues that comes into consideration is that there are two templates of weights to be built. The set of bottom-up weights z_{ij}^{bu}, each of which must store a real value belonging to the interval [0,1], and the set of top-down weights z_{ji}^{td}, each of which stores either the value '0' or '1'. The physical implementation of the bottom-up template memory presents the first hardware difficulty because the weights need either an analog or a digital memory with sufficient bits per weight so that the digital discretization does not affect the system performance. However, in the previous chapter, it could be seen from eq. (1.9)

$$z_{iJ}^{bu}(new) = \frac{Lz_{Ji}^{td}(new)}{L - 1 + |\mathbf{z}_J^{td}(new)|} \qquad (2.1)$$

that the bottom-up set $\{z_{ij}^{bu}\}$ and the top-down set $\{z_{ji}^{td}\}$ contain the same information: each of these sets can be fully computed by knowing the other set. The bottom-up set $\{z_{ij}^{bu}\}$ is a normalized version of the top-down set $\{z_{ji}^{td}\}$. Therefore, from a hardware implementation point of view, it would be desirable to implement physically only a binary valued set (one bit per weight) and introduce the normalization of the bottom-up weights during the computation of terms $\{T_j\}$. This way, the two sets $\{z_{ij}^{bu}\}$ and $\{z_{ji}^{td}\}$ can be substituted by a single binary valued set $\{z_{ij}\}$, and eq. (1.6)

$$T_j = \sum_{i=1}^{N} z_{iJ}^{bu} I_i = |\mathbf{z}_J^{bu} \cap \mathbf{I}|, \quad j = 1, \ldots, M \qquad (2.2)$$

modified to take into account the normalization effect of the original bottom-up weights,[1]

$$T_j = \frac{L|\mathbf{I} \cap \mathbf{z}_j|}{L - 1 + |\mathbf{z}_j|} \qquad (2.3)$$

---

[1] This type of modification is employed in the Fuzzy-ART model [Carpenter, 1991c].

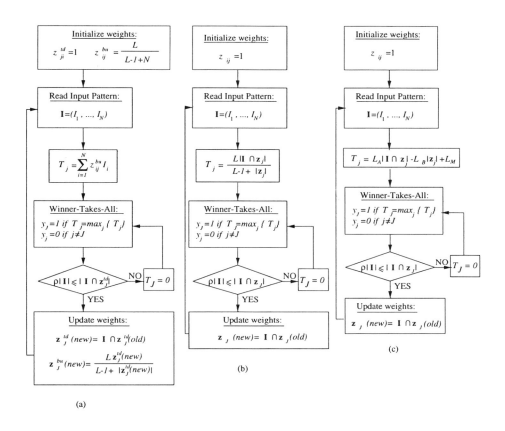

Figure 2.1. Different algorithmic implementations of the ART1 architecture: (a) original ART1 (b) ART1 with a single binary valued weight template (c) and VLSI-friendly ART1m

Considering this minor 'implementation modification', the original algorithm shown again in Fig. 2.1(a) would be transformed into that depicted in Fig. 2.1(b). The system level performance of the algorithms described by Figs. 2.1(a) and (b) is identical. There is no difference in the behavior between the two diagrams, and the one in Fig. 2.1(b) offers more attractive features from a hardware (as well as software) implementation point of view.

However, in Fig. 2.1(b), an extra division operation $T_j = L|\mathbf{I} \cap \mathbf{z}_j|/(L-1+|\mathbf{z}_j|)$ needs to be performed for each node in layer $F2$. This is an expensive hardware operation and would probably constitute a performance bottleneck in the overall system for both analog and digital circuit implementations. If possible, it would be very desirable to avoid this division operation.

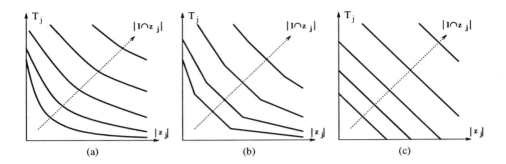

Figure 2.2. Illustration of the simplification process of the division operation: (a) original division operation, (b) piece-wise linear approximation, (c) linear approximation

Fig. 2.2(a) shows the curves that represent the division operation of eq. (2.3). A first simplification could be to substitute these curves by a piece-wise linear approximation as shown in Fig. 2.2(b). Such an approximation still presents some hardware difficulties and could also limit the performance of the overall system. A more drastic simplification would be to substitute the original operation by the operation represented by the set of curves of Fig. 2.2(c). Mathematically, the division operation has been substituted by a subtraction operation[2],

$$T_j = L_A |\mathbf{I} \cap \mathbf{z}_j| - L_B |\mathbf{z}_j| + L_M \qquad (2.4)$$

where L_A and L_B are positive parameters that play the role of the original L (and $L-1$) parameter. As shown in Appendix B, the condition $L_A > L_B$ must be imposed for proper system operation. $L_M > 0$ is a constant parameter needed[3] to ensure that $T_j \geq 0$ for all possible values of $|\mathbf{I} \cap \mathbf{z}_j|$ and $|\mathbf{z}_j|$.

Replacing a division operation with a subtraction one is a very important hardware simplification with significant performance improvement potential.

---

[2] Similar T_j functions (also called *distances* or *choice functions*) have been proposed by other authors for Fuzzy-ART [Carpenter, 1994]. Since ART1 can be considered a particular case of Fuzzy-ART when the input patterns are binary, Fuzzy-ART *choice functions* can also be used for ART1. These other *choice functions* also yield to ART1 architectures that preserve as well all the original computational properties [Serrano-Gotarredona, 1996a]. However, the *choice function* presented here is computationally less expensive and is easier to implement in hardware.
[3] In reality, parameter L_M has been introduced for hardware reasons [Serrano-Gotarredona, 1996b]. In a software implementation parameter L_M can be ignored.

Fig. 2.1(c) shows the final modified ART1 algorithm, which we will call throughout this Chapter the ART1m algorithm. In Appendix B we show that all the ART1 mathematical properties are preserved when replacing the division operation by a subtraction one. On the contrary, in the next Section we show that the result is a system with a slightly different input-output behavior, but nevertheless the parameters can be mapped from the original algorithm to the modified one to minimize the differences between the two.

It is worth mentioning here that substituting a division operation by a subtraction one means a significant performance boost from a hardware implementation point of view. Implementing physically division operators in hardware constraints significantly the whole system design and imposes limitations on the overall system performance.

In the case of analog hardware, there are ways to implement the division operation with compact dedicated circuits [Bult, 1987], [Sánchez-Sinencio, 1989], [Gilbert, 1990a], [Sheingold, 1976], but they usually suffer from low signal-to-noise ratios, limited signal range, noticeable distortion, or require exponential law devices (bipolar transistors [Roulston, 1990] or weak inversion MOS transistors [Tsividis, 1988]). In any case, the performance of the overall ART system would be limited by the lower performance of the division operators. If the division operators are eliminated the performance of the system would be limited by other simpler operators which, for the same VLSI technology, render considerable better performance figures. Furthermore, in the case of analog current mode signal processing the addition and subtraction of currents does not need any physical components, as we will see later in the Chapters on circuit implementations. Consequently, by eliminating the need of signal division, the circuitry is dramatically simplified and its performance drastically improved.

2.2 FUNCTIONAL DIFFERENCES BETWEEN ORIGINAL AND MODIFIED MODEL

As mentioned in Appendix B, the clustering behavior of ART1 and ART1m is governed by how all terms $\{T_j\}$ are initially ordered for a given input pattern **I**. It is easy to see that for the same input pattern **I** and weight templates $\{\mathbf{z}_j\}$, eqs. (2.3) and (2.4) will produce different values for terms T_j and their orderings will be different,

Original ART1 (Fig. 2.1(a)):
$$\frac{|\mathbf{I} \cap \mathbf{z}_{j_1}|}{L - 1 + |\mathbf{z}_{j_1}|} > \frac{|\mathbf{I} \cap \mathbf{z}_{j_2}|}{L - 1 + |\mathbf{z}_{j_2}|} > \frac{|\mathbf{I} \cap \mathbf{z}_{j_3}|}{L - 1 + |\mathbf{z}_{j_3}|}$$
(2.5)

Modified ART1 (Fig. 2.1(c)):
$$\frac{L_A}{L_B}|\mathbf{I} \cap \mathbf{z}_{l_1}| - |\mathbf{z}_{l_1}| > \frac{L_A}{L_B}|\mathbf{I} \cap \mathbf{z}_{l_2}| - |\mathbf{z}_{l_2}| > \ldots$$

where j_k might be different than l_k. The ordering resulting for the original ART1 description is modulated by parameter $L > 1$. For example, if $L - 1$ is very large compared to all $|\mathbf{z}_j|$ terms, then the ordering depends exclusively on the values of $|\mathbf{I} \cap \mathbf{z}_j|$,

$$|\mathbf{I} \cap \mathbf{z}_{j_1}| > |\mathbf{I} \cap \mathbf{z}_{j_2}| > |\mathbf{I} \cap \mathbf{z}_{j_3}| > \ldots \qquad (2.6)$$

If L is very close to 1, then the ordering depends on the ratios,

$$\frac{|\mathbf{I} \cap \mathbf{z}_{j_1}|}{|\mathbf{z}_{j_1}|} > \frac{|\mathbf{I} \cap \mathbf{z}_{j_2}|}{|\mathbf{z}_{j_2}|} > \frac{|\mathbf{I} \cap \mathbf{z}_{j_3}|}{|\mathbf{z}_{j_3}|} > \ldots \qquad (2.7)$$

Likewise, for the ART1m description, the ordering is modulated by a single parameter $\alpha = L_A/L_B > 1$. If α is extremely large, the situation in eq. (2.6) results. However, for α very close to 1, the ordering depends on the differences,

$$|\mathbf{I} \cap \mathbf{z}_{l_1}| - |\mathbf{z}_{l_1}| > |\mathbf{I} \cap \mathbf{z}_{l_2}| - |\mathbf{z}_{l_2}| > |\mathbf{I} \cap \mathbf{z}_{l_3}| - |\mathbf{z}_{l_3}| > \ldots \qquad (2.8)$$

Obviously, the behavior of the two ART1 descriptions will be identical for large values of L and α. However, moderate values of L and α are desired in practical ART1 applications. On the other hand, it can be expected that the behavior will also tend to be similar for very high values of ρ: if ρ is very close to 1, each training pattern will form an independent category. However, different training patterns will cluster into a shared category for smaller values of ρ. Therefore, a very similar behavior between ART1 and ART1m will be expected for high values of ρ, while more differences in behavior might be apparent for smaller values of ρ.

In order to compare the two algorithms' behavior, exhaustive simulations using randomly generated training patterns sets were performed[4]. As an illustration of a typical case where the two algorithms produce different learned templates, Fig. 2.3 shows the evolution of the memory templates, for both the ART1 and the ART1m algorithms, using a randomly generated training set of 10 patterns with 25 pixels each. Weight templates for original ART1 are named $\mathbf{z}_j$, while for ART1m they are named $\mathbf{z}'_j$. The vigilance parameter was set to $\rho = 0.4$. For the original ART1 $L = 5$, and for the ART1m $\alpha = 2$. In Fig. 2.3, boxed category templates are those that met the vigilance criterion and had the maximum T_j value. If the box is drawn with a continuous line, the corresponding $\mathbf{z}_j$ template suffered modifications due to learning. If the box is

---

[4]For all simulations in this Chapter, randomly generated training patterns sets were obtained with a 50 per cent probability for a pixel to be either '1' or '0'.

drawn with dashed line, learning did not alter the corresponding z_j template. Both algorithms stabilized their weights in 2 training trials. Looking at the learned templates we can see that input patterns 4 and 5 clustered in the same category for both algorithms (z_4 for original ART1 and z'_3 for ART1m). This also occurred for patterns 6 and 8 (z_3 and z'_2) and for patterns 3, 9 and 10 (z_5 and z'_5). However, patterns 1, 2, and 7 did not cluster in the same way in the two cases. In the original ART1 algorithm patterns 1 and 7 clustered into category z_1, while pattern 2 remained independent in category z_2. In the ART1m algorithm patterns 1 and 2 clustered together into category z'_1, while pattern 7 remained independent in category z'_4.

To measure a distance between two templates z_j and z'_j, let us use the Hamming distance between two binary patterns $\mathbf{a} \equiv (a_1, a_2, \ldots, a_N)$ and $\mathbf{b} \equiv (b_1, b_2, \ldots, b_N)$,

$$d(\mathbf{a}, \mathbf{b}) = \sum_{i=1}^{N} f_d(a_i, b_i), \tag{2.9}$$

where

$$f_d(a_i, b_i) = \begin{cases} 0 & \text{if } a_i = b_i \\ 1 & \text{if } a_i \neq b_i \end{cases} \tag{2.10}$$

We can use this metric to define the distance between two sets of patterns $\{z_j\}_{j=1}^{Q}$ and $\{z'_j\}_{j=1}^{Q}$ as that which minimizes

$$\sum_{i=1}^{Q} d(z_i, z'_{l_i}). \tag{2.11}$$

For this purpose, the optimal ordering of indexes $(l_1, l_2, \ldots, l_Q)$ must be found. In the case of Fig. 2.3 (where $Q = 5$), the distance D between the two learned patterns sets is given by,

$$D = d(z_1, z'_4) + d(z_2, z'_1) + d(z_3, z'_2) + d(z_4, z'_3) + d(z_5, z'_5) = 7, \tag{2.12}$$

In general, we can define the distance between two patterns sets $\mathbf{A} = \{\mathbf{a}_j\}_{j=1}^{Q}$ and $\mathbf{B} = \{\mathbf{b}_j\}_{j=1}^{Q}$ as,

$$D(\mathbf{A}, \mathbf{B}) = \min_{\{l_1, l_2, \ldots, l_Q\}} \left[\sum_{i=1}^{Q} d(\mathbf{a}_1, \mathbf{b}_{l_i}) \right] \tag{2.13}$$

46 ADAPTIVE RESONANCE THEORY MICROCHIPS

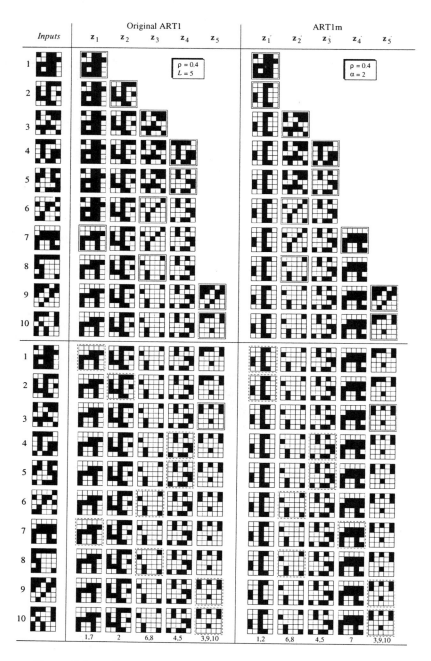

Figure 2.3. Comparative Learning Example ($\rho = 0.4, L = 5, \alpha = 2$)

In the case of Fig. 2.3, both algorithms produced the same number of learned categories. This does not always occur. For the case where a different number of categories results, we measured the distance between the two learned sets by adding as many uncommitted $F2$ nodes to the set with less categories as necessary to equal the number of categories. An uncommitted category has all its pixels set to '1'. Thus, having a different number of committed nodes drastically increases the resulting distance, and is consequently a strong penalty.

The simulation of Fig. 2.3 was repeated many times for different sets of randomly generated training patterns and sweeping the values of ρ, L, and α. For each combination of ρ, L, and α values, we repeated the simulation 100 times for different training patterns sets, and computed the average number of learned categories, learning trials, and distance between learned categories, as well as their corresponding standard deviations. Fig. 2.4 and Fig. 2.5 present the results of these simulations. Fig. 2.4(a) shows how the average number of learned categories changes with L (from 1.01 to 40) for different values of ρ, for the original ART1 algorithm. As ρ decreases, parameter L has more control on the average number of learned categories. Fig. 2.4(b) shows the standard deviation for the number of learned categories of Fig. 2.4(a). As the number of learned categories approaches the number of training patterns (10 in this case), standard deviation decreases. This happens for large values of L (independently of ρ) and for large values of ρ (independently of L). Fig. 2.4(c) and Fig. 2.4(d) show the same as Fig. 2.4(a) and Fig. 2.4(b) respectively, for the ART1m algorithm. As we can see, parameter α (swept from 1.01 to 5.0) of ART1m has more tuning power than parameter L of the original ART1. On the other hand, ART1m presents a slightly higher maximum standard deviation than the original ART1. Nevertheless, the qualitative behavior of both algorithms is similar. Fig. 2.4(e) and Fig. 2.4(f) show the average number of learning trials and their corresponding deviations, needed by the original ART1 algorithm to stabilize its learned weights. Fig. 2.4(g) and Fig. 2.4(h) show the same for the ART1m algorithm. As we can see, the ART1m algorithm needs a slightly higher average number of learning trials to stabilize. Also, the standard deviation observed for the ART1m algorithm is slightly higher. Finally, Fig. 2.5 shows the resulting average distances (as defined by eq. (2.13)) between learned categories of the ART1 and the ART1m algorithms. For ρ changing from 0.0 to 0.7 in steps of 0.1, each sub-figure in Fig. 2.5 depicts the resulting average distance for different values of L while sweeping α between 1.01 and 5.0.

It seems natural to expect that, for a given value of ρ and a given value of the original ART1 parameter L, there is an optimal value for the ART1m parameter α that will minimize the difference in behavior between the two algorithms. To find this relation between L and α for each ρ, we searched (for a given ρ and L) the value of α that minimizes the average distance between the learned patterns sets generated by the two algorithms. The results of these computations are shown in Fig. 2.6. Fig. 2.6(a) shows a family of curves (one for each value of ρ), that reveals the optimal value of α as a function of L. Fig. 2.6(b) shows the resulting minimum average distance between learned sets

48 ADAPTIVE RESONANCE THEORY MICROCHIPS

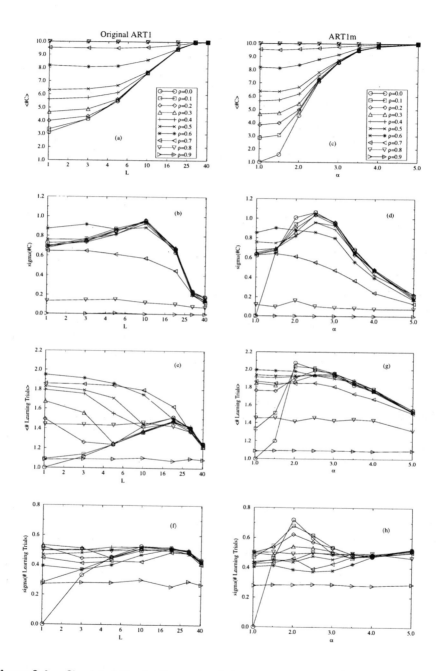

Figure 2.4. Simulated Results Comparing Behavior between ART1 and ART1m

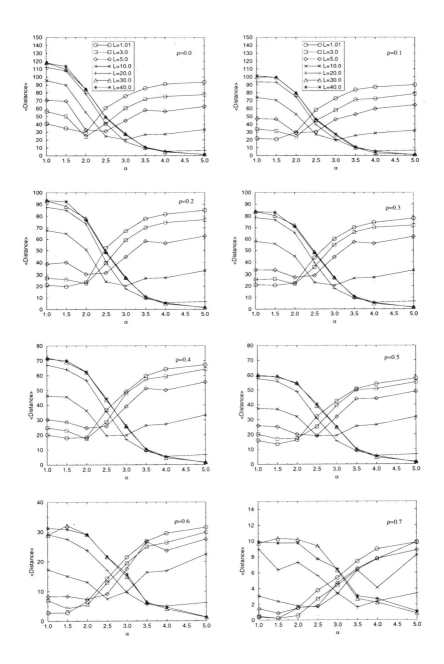

Figure 2.5. Learned Categories Average Distances

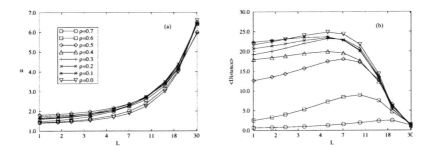

Figure 2.6. Optimal parameters fit between ART1 and ART1m

for the same family of curves. As shown in Fig. 2.6(a), the optimum fit between parameters α and L is very slightly dependent on the value of ρ.

As can be concluded from Figs.2.4, 2.5, 2.6, and the discussion in this Section, the behavior of the two algorithms is qualitatively the same although some slight quantitative differences can be observed. ART1m parameter α has a wider tuning range than original ART1 parameter L. On the other hand, ART1m needs a slightly higher number of learning trials than the original ART1. Also, there is an optimal adjustment between parameters α and L that minimizes the difference in behavior between the two algorithms, and this adjustment appears approximately independent of ρ.

A subtle difference between ART1 and ART1m is their behavior with respect to the empty input template ($\mathbf{I} \equiv \mathbf{0}$) or empty learned template ($\mathbf{z}_j \equiv \mathbf{0}$). The empty input pattern ($\mathbf{I} \equiv \mathbf{0}$) cannot be used in ART1, because

$$T_j(\mathbf{I} \equiv \mathbf{0}) = \frac{L|\mathbf{0} \cap \mathbf{z}_j|}{L - 1 + |\mathbf{z}_j|} = 0 \quad , \quad \forall j \tag{2.14}$$

so that $T_j = 0$ for all committed and uncommitted nodes. Layer $F2$ will not be able to pick a winner. However, for ART1m

$$T_j(\mathbf{I} \equiv \mathbf{0}) = L_A|\mathbf{0} \cap \mathbf{z}_j| - L_B|\mathbf{z}_j| + L_M = -L_B|\mathbf{z}_j| + L_M \tag{2.15}$$

and the $F2$ node with smallest $|\mathbf{z}_j|$ wins. This node automatically meets the vigilance criterion for any ρ,

$$\rho|\mathbf{I} \equiv \mathbf{0}| \leq |\mathbf{I} \cap \mathbf{z}_j| = 0 \tag{2.16}$$

On the other hand, assuming the empty template ($\mathbf{I} \equiv \mathbf{0}$) is never given as input, ART1 will never learn an empty category ($\mathbf{z}_j \equiv \mathbf{0}$) but ART1m could. Suppose a non-empty input pattern $\mathbf{I}$ is given such that the intersection $|\mathbf{I} \cap \mathbf{z}_j| = 0$ for all committed nodes j. Then for ART1 $T_j = 0$ for all committed nodes, but $T_j \neq 0$ for an uncommitted node. Consequently, a new category will be formed. For ART1m, if $|\mathbf{I} \cap \mathbf{z}_j| = 0$ for all committed nodes

$$T_j = -L_B |\mathbf{z}_j| + L_M \tag{2.17}$$

which might still be larger than the corresponding choice function for an uncommitted node

$$T_j = L_A |\mathbf{I}| - L_B N + L_M \tag{2.18}$$

However, the committed node (if selected before the uncommitted one) can only meet the vigilance criterion for $\rho = 0$

$$\rho |\mathbf{I}| \leq |\mathbf{I} \cap \mathbf{z}_j| = 0 \tag{2.19}$$

Therefore, the empty pattern is not valid for ART1 but it is a possible pattern for ART1m. In ART1, the empty pattern will never result as a consequence of training, but in ART1m if $\rho = 0$ a category template $\mathbf{z}_j \equiv \mathbf{0}$ might result even if no empty training pattern $\mathbf{I} \equiv \mathbf{0}$ was used.

In conclusion, there is a difference in behavior between ART1 and ART1m although both algorithms operate qualitatively in an equivalent manner. However, all the mathematical properties of algorithm ART1 are preserved for ART1m. In Appendix B all mathematical theorems available for the ART1 architecture [Carpenter, 1987] are formulated and proven for *Fast-Learning* ART1 and for *Fast-Learning* ART1m, one by one. Consequently, it is not clear to say which algorithm performs best from a functional point of view. But what is definitely clear is that ART1m (and its homologous for Fuzzy-ART) are much more advantageous from a hardware implementation point of view. This is the main point of view for this book and on which we will concentrate from here on.

3 ART1 AND ARTMAP VLSI CIRCUIT IMPLEMENTATION

3.1 INTRODUCTION

Two types of neural hardware engineers can be distinguished. The first designs "general purpose" hardware accelerators or systems that speed up neural algorithms running on conventional computers [Mauduit, 1992], [Jones, 1991], [Ramacher, 1991]. This kind of hardware allows considerable flexibility in the topology and operations of the neural systems. In this way algorithm researchers have a powerful tool to further develop neural algorithms and industry engineers have some attractive chips that significantly speed up their neural commercial products. The second type of hardware engineers are those who design a real-time system for a specific application. They must select the best-suited algorithm and map it into hardware. This achieves a close-to-optimum efficient hardware for a limited range of applications. The work described in this book falls into this second category of hardware engineering. The specific application is real-time clustering of input patterns.

A clustering device is a device able to build categories from a collection of patterns. A real-time clustering device has to be able to do this at the speed of arrival of the patterns. There are some clustering algorithms [Kohonen, 1989], [Bezdek, 1981], [Duda, 1973], [Hartigan, 1975], [Dubes, 1988], [Pao, 1989] that need to be trained off-line to build the categories. For a real-time clustering device, however, it would be desirable to use an algorithm that can be trained on-line: if a new pattern arrives the algorithm updates its internal knowledge (instead of erasing all the accumulated knowledge and retrain with the old and new collection of patterns).

For the second type of neural hardware engineers, the issue of efficiently implementing in hardware a real size neural network is not a trivial task. Many neural network algorithms are available in the literature which have been developed, studied, and optimized for applications through computer and/or software based systems. Consequently, when designing a hardware realization, engineers face many problems like excessive interconnectivity, high resolution of weights, high precision of operations, complicated operator requirements (e.g., integrals and derivatives), high number of neurons required for a real-world application, etc. Many times some of these requirements can be relaxed, the topology modified, or the operations simplified, with no significant deterioration of global operation of the neural system but with a considerable boost in the hardware performance. Modifying neural algorithms to make them more VLSI-friendly and produce more efficient hardware should be a common practice among neural hardware engineers of the second type [Rodríguez-Vázquez, 1993], [Linares-Barranco, 1993]. After selecting an appropriate neural algorithm the next step consists of studying how far the algorithm can be simplified without performance degradation. The simplifications have to be hardware-oriented, so that the final combination of "theoretical algorithm" + "hardware circuit technique" results in a high performance real time system. The success of the hardware system depends on the selection of the algorithm, the selection of a powerful circuit design technique, and how the algorithm is modified to efficiently "marry" the circuit technique resulting in an optimum performance final system.

In this book Adaptive Resonance Theory algorithms are considered mainly due to their attractive hardware-oriented properties, as well as the theoretical computational properties. In this Chapter we will concentrate on the implementation of ART1 based systems (i.e., ART1 and ARTMAP) which work on binary valued input patterns. In later Chapters we will extend to algorithms for analog valued input patterns. As described in the previous Chapter, for the case of ART1 we also chose to slightly modify the mathematical algorithm to obtain more efficient hardware. This modification allows the use of simpler operations while preserving all the computational properties of the original ART1 architecture. As an extra bonus, the hardware circuit introduces a significant speed improvement as it automatically parallels the sequential ART search process (checking of the vigilance criterion). The advantageous features of the ART1 algorithm are described next.

3.2 HARDWARE-ORIENTED ATTRACTIVE PROPERTIES OF THE ART1 ALGORITHM

Fast Learning

In performance comparison of hardware implementations, a common figure of merit is the number of interconnections per second. More refined figures have been proposed that include resolution and precision [Keulen, 1994]. However, these figures would be reasonably fair criteria for the first type of hardware engineering mentioned above, the general-purpose one. In order to compare hardware systems of the second type, the specific-application neural hardware, some global figure must be used that evaluates the overall system performance. Usually this figure will be application dependent. In this Chapter, since we are concerned with real-time clustering of binary input patterns, an appropriate figure of merit might be

$$ppc/s = \frac{\text{number of \textbf{p}atterns processed}}{seconds} \times \textbf{p}ixels \times categories \quad (3.1)$$

where,

- **number of patterns processed/second** is the speed at which patterns are classified and learned (including the number of learning trials required). This speed generally depends on the patterns themselves, and on the knowledge already stored in the system. Therefore, this speed can be given as an average or as the slowest case measured.
- **pixels** is the maximum number of pixels of the input patterns.
- **categories** is the maximum number of categories the system is able to form.

As we will see later in "Experimental Results", one of the chips is able to cluster up to 18 different categories of binary patterns with 100 pixels, while classifying and learning each pattern in less than $1.8\mu s$. Since ART1 learns on-line, 1 iteration of input patterns presentations provides the system with sufficient knowledge to perform properly. This results in a *ppc/s* of

$$ppc/s = \frac{n \text{ patterns}}{1 \text{ iteration} \times n \text{ patterns} \times 1.8\mu s} \times 100 \text{ pixels} \times 18 \text{ categories} = \quad (3.2)$$
$$= 1.0 \times 10^9 ppc/s$$

If we would like to obtain the same performance using Backpropagation based hardware, and assuming the network would learn with 10,000 iterations of patterns presentations, this means that a speed of $180ps$ would be needed for each pattern classification and corresponding weights update. Assuming this task could be performed with a Backpropagation network with 100 input neurons, 5 hidden-layer neurons, and 5 output neurons[1] (which means a total of $100 \times 5 + 5 \times 5 = 525$ interconnections), and that the speed of feedforward classification is the same as for feedback learning, hardware able to perform

$$\frac{2 \times 525 \text{ connections}}{180 ps} = 5.83 \times 10^{12} \text{ connections/s + connection-updates/s} \tag{3.3}$$

would be needed. For the chips described in this Chapter, since they are based on the powerful ART1 algorithm, the above performance can be achieved with a hardware of only 4.4×10^9 *connections/s* plus *connections-updates/s*.

Note that the Backpropagation algorithm is not appropriate for clustering applications, and comparing it against ART1 is slightly unfair. There are other algorithms available in the literature that have been developed specially for clustering applications [Kohonen, 1989], [Bezdek, 1981], [Duda, 1973], [Hartigan, 1975], [Dubes, 1988], [Pao, 1989]. However, they usually do not provide all the computational properties mentioned in Chapter 1 and Appendix B, specially the *"On-Line Learning"* property which is crucial for real-time clustering, or they present serious difficulties when trying to map them into hardware.

Binary Memory

Another hardware attractive feature that an ART1 based implementation offers with respect to others, is that the interconnection weights do not have to be analog, as discussed in Chapter 2. Most of the neural algorithms reported in the literature require a real-valued set of weights defined within a certain interval. These weights can be discreetized in a number of digital steps, but the granularity required for proper operation of the system is usually very fine (around 16-bits for the Back-Propagation algorithm [Riedmiller, 1994]). Even worse, in some cases the granularity requirements become more severe as the size of the system increases. For example, in a BAM system [Kosko, 1987] of $N \times M$ neurons, storage capacity has been heuristically estimated to be around $n_p = (N \times M)^{1/4}$ [Wand, 1990], where n_p is the average maximum number of patterns that can be stored. The resolution required by the interconnection weights in this case is at least $n_p + 1$. For chips based on the ART1 algorithm, since they require only binary-valued weights, the resolution of the weights is not affected by the size nor the storage capacity of the system. This, and the

---

[1] Optimistically, a Backpropagation network with 5 output neurons could code up to 32 categories

non necessity of analog weights is one of the most hardware attractive features of the ART1 algorithm.

Scaling

Another consideration to take into account during the design of a hardware system is how it scales up with size and performance. We have already mentioned that some neural systems need to increase their weight resolution as they scale up. Another feature is how their size and interconnectivity scale up with pattern size or storage capacity. For an ART1 based system, the number of neurons N in the bottom layer is the number of pixels of the patterns, the number of neurons M in the top layer is the maximum number of categories, and $N \times M$ is the number of synapses. Therefore, this system scales up linearly with storage capacity (M) and input pixels (N). For a BAM system, for example, the size scales quadratically with the storage capacity and the number of pixels.

In the case of an analog hardware, random and systematic errors due to fabrication process variations will appear. A neural network can usually cope very well with random errors, even if the size of the system increases. However, systematic errors may accumulate as the system increases and may render the complete network useless as it scales up. For example, if using transconductors based circuits (T-mode) [Linares-Barranco, 1993], [Linares-Barranco, 1992], due to the limited input voltage range of the transconductors (around $100mV$ if they have to be of minimum area) the practical maximum number of synapses is very limited (if each synapse introduces $1mV$ of systematic offset, 50 synapses would completely saturate the transconductor but misfunctions would appear with less synapses). The chosen circuit technique must be either insensitive to the accumulation of systematic errors, or allow for some kind of calibration technique to overcome them. In ART1 if there is a systematic (common) component being added to the synaptic current sources L_A and/or L_B, there will be a systematic (common) component affecting all T_j terms. Since terms T_j have to compete in the WTA, a systematic component will not affect the selection of the winners (as long as the WTA does not saturate, but WTA circuits can allow very wide ranges of input signals).

3.3 CIRCUIT DESCRIPTION

Fig. 2.1(c) shows the VLSI-friendly algorithmic ART1 description, discussed in the previous Chapter, which we want to map into hardware.

The operations in Fig. 2.1(c) that need to be implemented are the following:

- Generation of the terms T_j or "choice functions". Since z_{ij} and I_i are binary valued (0 or 1), "binary multiplication" and analog addition/subtraction operations are required.

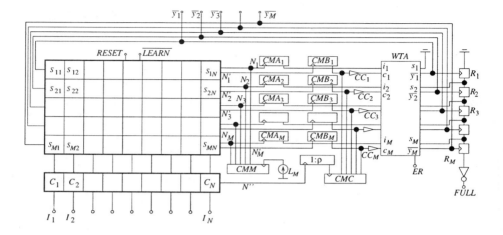

Figure 3.1. Hardware Block Diagram for the ART1m Algorithm

- Winner-Takes-All (WTA) operation to select the maximum T_j term.
- Comparison of the term $\rho|\mathbf{I}|$ with $|\mathbf{I} \cap \mathbf{z}_J|$.
- Deselection of terms T_J for which $\rho|\mathbf{I}| > |\mathbf{I} \cap \mathbf{z}_J|$.
- Update of weights.

The first three operations require a certain amount of precision (they are analog operations), while the last two operations are not precise (they are digital operations). A reasonable precision for a massive parallel analog circuit implementation is around 1 or 2% (equivalent to 6-bits), while handling input patterns with N of the order of hundreds. Fig. 3.1 shows a possible hardware block diagram that would physically implement the algorithm of Fig. 2.1(c). The circuit consists of an $M \times N$ array of synapses $S_{11}, S_{12}, \ldots, S_{MN}$, a $1 \times N$ array of controlled current sources $C_1, C_2, \ldots C_N$, two $1 \times M$ arrays of unity-gain current mirrors $CMA_1, \ldots, CMA_M, CMB_1, \ldots, CMB_M$, a $1 \times M$ array of current comparators $CC_1, \ldots, CC_M$, an M-input WTA circuit, two M-output unity-gain current mirrors CMM and CMC, and an adjustable-gain ($0 < \rho \leq 1$) current mirror. Registers $R_1, \ldots, R_M$ are optional, and their function is explained later.

Each synapse receives two input signals $\overline{y_j}$ and I_i, has two global control signals $RESET$ and $\overline{LEARN}$, stores the value of z_{ij}, and generates two output currents:

- the first goes to the input of current mirror CMA_j (node N_j) and is $L_A z_{ij} I_i - L_B z_{ij}$.

- the second goes to the input of current mirror CMB_j (node N'_j) and is $L_A z_{ij} I_i$.

All synapses in the same row j $(S_{j1}, S_{j2}, \ldots, S_{jN})$ share the two nodes (N_j and N'_j) into which the currents they generate are injected. Therefore, the input of current mirror CMA_j receives the current

$$T_j = L_A \sum_{i=1}^{N} z_{ij} I_i - L_B \sum_{i=1}^{N} z_{ij} + L_M = L_A |\mathbf{I} \cap \mathbf{z}_j| - L_B |\mathbf{z}_j| + L_M \quad (3.4)$$

while the input of current mirror CMB_j receives the current

$$L_A \sum_{i=1}^{N} z_{ij} I_i = L_A |\mathbf{I} \cap \mathbf{z}_j| \quad (3.5)$$

Current L_M, which is replicated M times by current mirror CMM has an arbitrary value as long as it assures that all terms T_j are positive.

Each element of the array of controlled current sources C_i has one binary input signal I_i and generates the current $L_A I_i$. All elements C_i share their output node N'', so that the total current they generate is $L_A |\mathbf{I}|$. This current reaches the input of the adjustable gain ρ current mirror, and is later replicated M times by current mirror CMC.

Each of the M current comparators CC_j receives the current $L_A |\mathbf{I} \cap \mathbf{z}_j| - L_A \rho |\mathbf{I}|$ and compares it against zero (i.e. it checks the vigilance criterion for category j). If this current is positive, the output of the current comparator falls, but if the current is negative the output rises. Each current comparator CC_j output controls input c_j of the WTA. If c_j is high the current sunk by the WTA input i_j (which is T_j) will not compete for the winning node. On the contrary, if c_j is low, input current T_j will enter the WTA competition. The outputs of the WTA $\overline{y_j}$ are all high, except for that which receives the largest $\overline{c_j} T_j$: such output, denominated $\overline{y_J}$, will fall.

Now we can describe how the operation of the circuit in Fig. 3.1 follows that of the flow diagram of Fig. 2.1(c). All synaptic memory values z_{ij} are initially set to '1' by the $RESET$ signal. Once the input vector $\mathbf{I}$ is activated, the M rows of synapses generate the currents $L_A |\mathbf{I} \cap \mathbf{z}_j| - L_B |\mathbf{z}_j|$ and $L_A |\mathbf{I} \cap \mathbf{z}_j|$, and the row of controlled current sources $C_1, \ldots, C_N$ generates the current $L_A |\mathbf{I}|$. Each current comparator CC_j will prevent current $T_j = L_A |\mathbf{I} \cap \mathbf{z}_j| - L_B |\mathbf{z}_j| + L_M$ from competing in the WTA if $\rho |\mathbf{I}| > |\mathbf{I} \cap \mathbf{z}_j|$. Therefore, the effective WTA inputs are $\{\overline{c_j} T_j\}$, from which the WTA chooses the maximum, making the corresponding output $\overline{y_J}$ fall. Once $\overline{y_J}$ falls, the synaptic control signal $\overline{LEARN}$ is temporarily set low, and all z_{iJ} values will change from '1' to 'I_i'.

Note that initially (when all $z_{ij} = 1$),

$$\overline{c_j}T_j = L_A|\mathbf{I}| - L_B N + L_M \quad \forall j \tag{3.6}$$

This means that the winner will be chosen among M equal competing inputs, basing the election on mismatches due to random process parameter variations of the transistors. Even after some categories are learned, there will be a number of uncommitted rows ($z_{1j} = \cdots = z_{Nj} = 1$) that generate the same competing current of eq. (3.6). The operation of a WTA circuit in which there are more than 1 equal and winning inputs becomes more difficult and in the best case, renders slower operation. To avoid these problems M D-registers, $R_1, \ldots, R_M$, might be added. Initially these registers are set to '1' so that the WTA inputs $s_2, \ldots, s_M$ are high. Inputs $s_1, \ldots, s_M$ have the same effect as inputs $c_1, \ldots, c_M$: if s_j is high T_j does not compete for the winner, but if s_j is low T_j enters the WTA competition. Therefore, initially only $\overline{c_1}T_1$ competes for the winner. As soon as $\overline{y_1}$ falls once, the input of register R_1 (which is '0') is transmitted to its output making $s_2 = 0$. Now both $\overline{c_1}T_1$ and $\overline{c_2}T_2$ will compete for the winner. As soon as $\overline{c_2}T_2$ wins once, the input of register R_2 is transmitted to its output making $s_3 = 0$. Now $\overline{c_1}T_1$, $\overline{c_2}T_2$, and $\overline{c_3}T_3$ will compete, and so on. If all available $F2$ nodes ($y_1, \ldots, y_M$) have won once, the $FULL$ signal rises, advising that all $F2$ nodes are storing a category. The WTA control signal ER enables operation of these registers. Next we describe the circuit implementation for each block in Fig. 3.1.

Synaptic Circuit and Controlled Current Sources

The details of a synapse S_{ji} are shown in Fig. 3.2(a). It consists of three current sources (two of value L_A and one of value L_B), a two-inverter loop (acting as a Flip-Flop), and nine MOS transistors working as switches. As can be seen in Fig. 3.2(a) each synapse generates the currents $L_A z_{ij} I_i - L_B z_{ij}$ and $L_A z_{ij} I_i$. The $RESET$ control signal sets z_{ij} to '1'. Learning is performed by making z_{ij} change from '1' to '0' whenever $\overline{LEARN} = 0$, $\overline{y_j} = 0$, and $I_i = 0$.

Fig. 3.2(b) shows the details of each controlled current switch C_i. If $I_i = 0$ no current is generated, while if $I_i = 1$, the current L_A is provided.

Winner-Takes-All (WTA) Circuit

Fig. 3.3 shows the details of the WTA circuit. It is based on Lazzaro's WTA [Lazzaro, 1989], which consists of the array of transistors MA and MB, and the current source I_{BIAS}. Transistor MC has been added to introduce a cascode effect and increase the gain of each cell. Transistors MX, MY, and MZ transform the output current into a voltage, which is then inverted to generate $\overline{y_j}$. Transistor MT disables the cell if c_j is high, so that the input current T_j

ART1 AND ARTMAP VLSI CIRCUIT IMPLEMENTATION   61

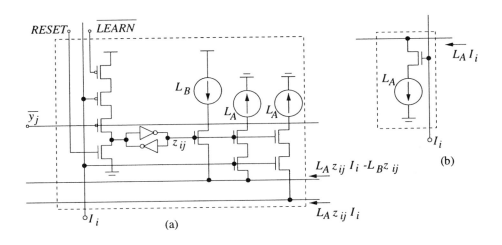

Figure 3.2. (a) Details of Synaptic Circuit S_{ij}. (b) Details of Controlled Current Source Circuit C_i

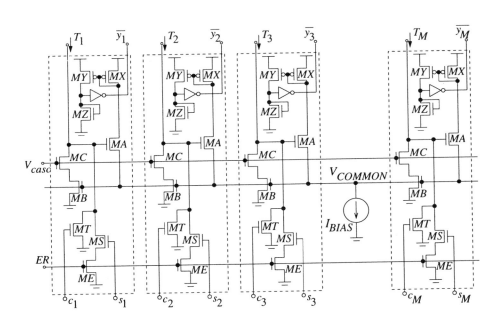

Figure 3.3. Circuit Schematic of Winner-Takes-All (WTA) Circuit

62 ADAPTIVE RESONANCE THEORY MICROCHIPS

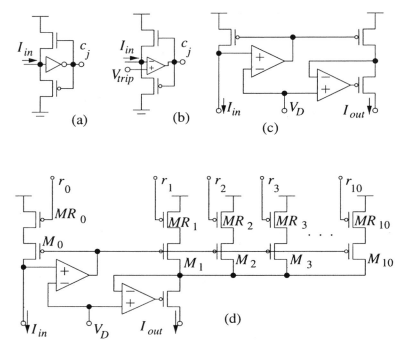

Figure 3.4. (a) Circuit Schematic of Current Comparator using Digital Inverter or (b) OTA, (c) of Active-Input Regulated-Cascode Current Mirror, (d) and of Adjustable Gain ρ Current Mirror.

will not compete for the winner. Transistors MS and ME have the same effect as transistor MT: if signals ER and s_j are high, T_j will not compete.

Current Comparators

The circuit used for the current comparators is shown in Fig. 3.4(a). Such a comparator forces an input voltage approximately equal to the inverters trip voltage, has extremely high resolution (less than $1pA$), and can be extremely fast (in the order of $10-20ns$ for input around $10\mu A$) [Rodríguez-Vázquez, 1995]. Note that in the steady state the inverter is biased in its trip-point, and therefore is consuming a significant amount of current. If power consumption is of concern this is not acceptable. Instead, the comparator of Fig. 3.4(b) should be used which uses an OTA [Mead, 1989] as voltage amplifier.

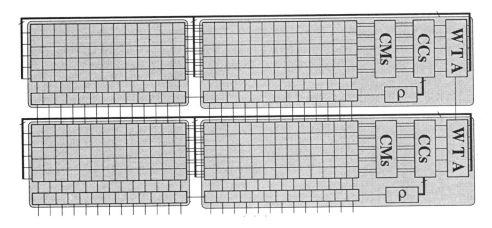

Figure 3.5. Interchip Connectivity for Modular System Expansion

Current Mirrors

Current Mirrors CMA_j, CMB_j, CMM, CMC, and the ρ-gain mirror have been laid out using common centroid layout techniques to minimize matching errors and keep a 6-bit precision for the overall system. For current mirrors CMA_j and CMB_j a special topology has been used, shown in Fig. 3.4(c) [Serrano-Gotarredona, 1994]. This topology forces a constant voltage V_D at its input and output nodes, thus producing a virtual ground in the output nodes of all synapses, which reduces channel length modulation distortion improving matching between the currents generated by all synapses. In addition, the topology of Fig. 3.4(c) presents a very wide current range with small systematic (non-random) matching errors [Serrano-Gotarredona, 1994].

The adjustable gain ρ current mirror also uses this topology, as shown in Fig. 3.4(d). Transistor M_0 has a geometry factor (W/L) 10 times larger than transistors $M_1, \ldots, M_{10}$. Transistors $MR_1, \ldots, MR_{10}$ act as switches (controlled by signals $r_1, \ldots, r_{10}$), so that the gain of the current mirror can be adjusted between $\rho = 0.0$ to $\rho = 1.0$ in steps of 0.1, while maintaining $r_0 = 0$. By making r_0 slightly higher than 0 Volts, ρ can be fine tuned.

3.4 MODULAR SYSTEM EXPANSIVITY

The circuit of Fig. 3.1 can be expanded both horizontally, increasing the number of input patterns from N to $N \times n_H$, and vertically increasing the number of possible categories from M to $M \times n_V$, just by assembling an array of $n_H \times n_V$ chips. Fig. 3.5 shows schematically the interconnectivity between chips in the case of a 2×2 array.

Vertical expansion of the system is possible by making several chips share the input vector terminals $I_1, \ldots, I_N$, and node V_{COMMON} of the WTA (see Fig. 3.3). Thus, the only requirement is that V_{COMMON} be externally accessible. Horizontal expansion is directly possible by making all chips in the same row share their N_j, N'_j, and N'' nodes, and isolating all except one of them, from the current mirrors CMA_j, CMB_j, and the adjustable gain ρ-mirror. Also, all synapse inputs $\overline{y_j}$ must be shared.

Both vertical and horizontal expansion degrades the system performance. Vertical expansion causes degradation because the WTA becomes distributed among several chips. For the WTA of Fig. 3.3, all MA transistors must match well, which is very unlikely if they are in different chips. A solution for this problem is to use a WTA circuit technique based on current processing and replication, insensitive to inter-chip transistor mismatches [Serrano-Gotarredona, 1995], [Serrano-Gotarredona, 1998a], which is described in Chapter 4.

Horizontal expansion degrades the performance because current levels have to be changed:

- Either currents L_A and L_B are maintained the same, which makes the current mirrors CMA_j, CMB_j, CMM, $1 : \rho$, and CMC, the current comparators CCj, and the WTA to handle higher currents. This may cause malfunctioning due to eventual saturation in some of the blocks.

- Or currents L_A and L_B are scaled down so that the current mirrors CMA_j, CMB_j, CMM, $1 : \rho$, and CMC, the current comparators CCj, and the WTA handle the same current level. However, this degrades the matching among synaptic current sources [Pelgrom, 1989].

Also, horizontal expansion requires to access externally $3M$ lines for each chip (nodes $\overline{y_j}$, N_j, and N'_j). If M is large (of the order of hundreds) horizontal expansion becomes unfeasible.

3.5 IMPLEMENTATION OF SYNAPTIC CURRENT SOURCES

The physical implementation of the *Synaptic Current Sources* is the one that most critically affects the final density of the resulting system as well as its precision.

The most simple way would be to build a current mirror with multiple outputs: for example, for current sources L_B, one can make a current mirror with $N \times M$ output transistors and have each of them placed in a synapse. For the L_A current sources a mirror with $2N \times M$ outputs would be needed. The problem with this implementation is that the distance between input and output transistors can become extremely large and therefore produce a very large mismatch between input and output currents. For example, if we want to make an array of size around $1cm^2$ using $40\mu m \times 40\mu m$ transistors, biased

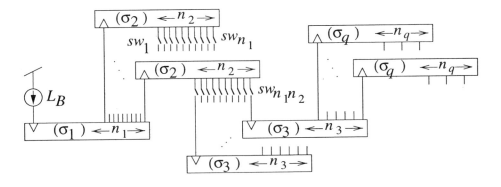

Figure 3.6. Cascade of Current Mirrors for Low Mismatch

with currents of $10\mu A$, a worst case mismatch of 10% would be optimistic. However, if instead of distances of the order of $1cm$, we can spread an array of transistors over distances of the order of $1mm$, we might be able to achieve mismatches below 1%. Appendix C explains a cheap and easy methodology for characterizing transistor mismatch of a given technology. This methodology also allows us to identify approximately what is the maximum die area over which we can spread an array of transistors (like the outputs of a current mirror) while maintaining mismatch below or close to 1%.

As opposed to large transistor arrays, a single current mirror, with a reduced number of outputs (like 10), a reasonable transistor size (like $40\mu m \times 40\mu m$), a moderate current (around $10\mu A$), and using common centroid layout techniques can be expected to have a mismatch error standard deviation σ_q of less than 1% [Pelgrom, 1989], [Appendix C]. By cascading several of these current mirrors in a tree-like fashion as is shown in Fig. 3.6 (for current sources L_B), a high number of current sources (copied from a single common reference) can be generated with a mismatch equal to

$$\sigma_{Total}^2 = \sigma_1^2 + \sigma_2^2 + \cdots + \sigma_q^2 \tag{3.7}$$

Each current mirror stage introduces a random error of standard deviation σ_k. This error can be reduced by increasing the transistor areas of the current mirrors. Since the last stage q has a higher number of current mirrors, it is important to keep their area low. For previous stages the transistors can be made larger to contribute with a smaller σ_k, because they are less in number and will not contribute significantly to the total transistor area. For current sources L_A, a circuit similar to that shown in Fig. 3.6 can be used. Current L_B in Fig. 3.6 (and similarly current L_A) is injected externally into the chip so that parameter $\alpha = L_A/L_B$ can be controlled.

In the next Sections we provide experimental results related to two ART1 chip implementations. They differ only in the way the L_A and L_B current sources are implemented. In the first prototype they were implemented using the current mirror tree approach of [2] Fig. 3.6. This produces a very high area consumption (since each stage, including the last, is laid out using common centroid structures). We were able to fit an array of 100×18 synapses into a $1 cm^2$ die, for a $1.6 \mu m$ CMOS technology.

The second prototype is made by directly making one current mirror with $N \times M$ outputs (for L_B) and another with $2N \times M$ outputs (for L_A), but limiting the array to $2.75 mm^2$. We were able to fit an array of 50×10 synapses into this area, for a $1.0 \mu m$ CMOS technology.

The optimum strategy for building large arrays would be to tile a large area (like $1 cm^2$) into pieces of smaller areas (like $1 - 2 mm^2$). In each tile one can implement the current mirrors with a very large number of outputs (as in the approach of the second prototype), and feed the input of each tile current mirror using a tree-mirror structure (as in the first prototype). This approach is schematically depicted in Fig. 3.7. When drawing the layout, care should be taken to assure that each synapse has a similar neighborhood whether or not it lies on the edge of a tile or not. This would avoid matching degradation due to layout neighborhood effects [Gregor, 1992].

Weights Read Out

A very important feature for test purposes is to be able to read out the weight z_{ij} for each synapse in the array. In the first prototype, the switches sw_1 to $sw_{n_1 \times n_2}$ of Fig. 3.6 were added to enable z_{ij} read out column by column and test the progress of the learning algorithm. These switches are all ON during normal operation of the system. However, for weights read-out, all except one will be OFF. The switch that is ON is selected by a decoder inside the chip, so that only column i of the synaptic array of Fig. 3.1 injects the current $z_{ij} L_B$ to nodes N_j. During weights read out, either current L_A is set to zero or input pattern $\mathbf{I} \equiv \mathbf{0}$ is loaded, so that only the L_B current sources inject current into nodes N_j. All nodes N_j can be isolated from current mirrors CMA_j, and connected to output pads to sense the currents $z_{ij} L_B$.

For the second prototype an extra MOS switch was added in series with each current source L_A and L_B. This switch is controlled by a common line for each column of the synaptic array. During normal operation all these switches are ON, but during weights read out all are OFF except for one column. The column is selected through an on-chip decoder.

---

[2] The chip was designed in this way because at that time no information on transistor mismatch over large distances ($1 cm$) was available.

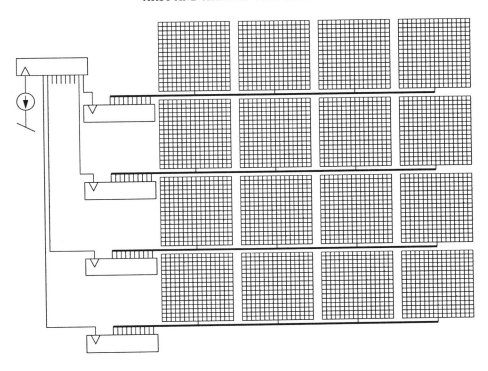

Figure 3.7. Optimum Strategy for Minimum Current Sources Mismatch for Large Area Arrays

3.6 EXPERIMENTAL RESULTS OF FIRST PROTOTYPE

Following the approach of Fig. 3.6 a prototype chip was fabricated in a standard single-poly double-metal $1.6\mu m$ CMOS digital process (Eurochip ES2). The die area is $1cm^2$ and it has been mounted in a 120-pin PGA package. This chip implements an ART1 system with 100 nodes in the $F1$ layer and 18 nodes in the $F2$ layer. Most of the pins are intended for test and characterization purposes. All the subcircuits in the chip can be isolated from the rest and conveniently characterized. The $F1$ input vector **I**, which has 100 components, has to be loaded serially through one of the pins into a shift register. The time delay measurements reported next do not include the time for loading this shift register.

The experimental measurements provided in this Section have been divided into four parts. The first describes DC characterization results of the elements that contribute critically to the overall system precision. These elements are the WTA circuit and the synaptic current sources. The second describes time delay measurements that contribute to the global throughput time of the system. The

Table 3.1. Precision of the WTA

T_j	$10\mu A$	$100\mu A$	$1mA$
$\sigma(T_j)$	1.73%	0.86%	0.99%

third presents system level experimental behaviors obtained with digital test equipment (HP82000). Finally, the fourth focuses on yield and fault tolerance characterizations.

System Precision Characterizations

The ART1 chip was intended to achieve an equivalent 6-bit (around 1.5% error) precision. The part of the system that is responsible for the overall precision is formed by the components that perform analog computations. These components are (see Fig. 3.1) all current sources L_A and L_B, all current mirrors CMA_j, CMB_j, CMM, $1:\rho$, CMC, the current comparators CC_j, and the WTA circuit. The most critical of these components (in precision) is the WTA circuit. Current sources and current mirrors can be made to have mismatch errors below 0.2% [Pelgrom, 1989], [Lakshmikumar, 1986], [Michael, 1992], [Shyu, 1984], [Appendix C], at the expense of increasing transistors area and current, decreasing distances between matched devices, and using common centroid layout techniques [Allen, 1987]. This is feasible for current mirrors CMA_j, CMB_j, CMM, $1:\rho$, and CMC, which appear in small numbers. However, the area and current level is limited for the synaptic current sources L_A and L_B, since there are many of them. Therefore, WTA and current sources L_A and L_B are the elements that limit the precision of the overall system, and their characterization results will be described next.

WTA Precision Measurements. L_A and L_B will have current values of $10\mu A$ or less. The maximum current a WTA input branch can receive is (see eq. (3.4)),

$$T_j|_{max} = L_M + \left[\sum_{i=1}^{100} z_{ij}(L_A I_i - L_B)\right]_{max} = L_M + 100(L_A - L_B) \quad (3.8)$$

which corresponds to the case where all z_{ij} and I_i values are equal to '1' (remember that $L_A > L_B > 0$). In our circuit the WTA was designed to handle input currents of up to $1.5mA$ for each input branch. In order to measure the precision of the WTA, all input currents except two were set to zero. Of these

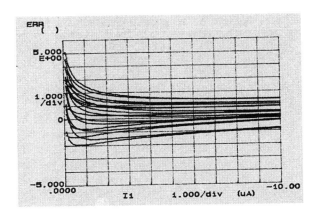

Figure 3.8. Measured Mismatch Error (in %) between 18 arbitrary L_A Current Sources

two inputs one was set to $100\mu A$ and the other was swept between $98\mu A$ and $102\mu A$. This will cause their corresponding output voltages $\overline{y_j}$ to indicate an interchange of winners. The transitions do not occur exactly at $100\mu A$. Moreover, the transitions change with the input branches. The standard deviation of these transitions was measured as $\sigma = 0.86\mu A$ (or 0.86%). Table 3.1 shows the standard deviation (in %) measured when the constant current is set to $10\mu A$, $100\mu A$, and $1mA$.

Synaptic Current Sources Precision Measurements. The second critical precision error source of the system is the mismatch between synaptic current sources. In the chip each of the $N \times (2M+1)$ L_A current sources and each of the $N \times M$ L_B current sources could be isolated and independently characterized. Fig. 3.8 shows the measured mismatch error (in %) for 18 arbitrary L_A current sources when sweeping their currents between $0.1\mu A$ and $10\mu A$. As can be seen in Fig. 3.8, for currents higher than $5\mu A$ the standard deviation of the mismatch error is below 1%. The same result is obtained for the L_B current sources.

Throughput Time Measurements

For a real-time clustering device the throughput time can be defined as the time needed for each input pattern to be processed. During this time the input pattern has to be classified into one of the pre-existing categories or assigned to a new one, and the pre-existing knowledge of the system has to be updated to incorporate the new information the input pattern carries. From a circuit point of view, this translates into the measurement of two delay times:

Table 3.2. Delay Times of the WTA

T_1^a	T_1^b	T_2	$T_3, \ldots T_{18}$	t_{d1}	t_{d2}
$0\mu A$	$200\mu A$	$100\mu A$	0	$550ns$	$570ns$
$0\mu A$	$1mA$	$500\mu A$	0	$210ns$	$460ns$
$100\mu A$	$150\mu A$	$125\mu A$	$100\mu A$	$660ns$	$470ns$
$400\mu A$	$600\mu A$	$500\mu A$	$400\mu A$	$440ns$	$400ns$
$500\mu A$	$1.50mA$	$1.00mA$	$500\mu A$	$230ns$	$320ns$
$90\mu A$	$110\mu A$	$100\mu A$	0	$1.12\mu s$	$1.11\mu s$
$490\mu A$	$510\mu A$	$500\mu A$	0	$1.19\mu s$	$1.06\mu s$
$990\mu A$	$1.01mA$	$1.00mA$	0	$380ns$	$920ns$

1. The time needed by the WTA to select the maximum among all $\overline{c_j}T_j$.

2. The time needed by the synaptic cells to change z_{ij} from its old value to $y_j I_i z_{ij}$.

WTA Delay Measurements. The delay introduced by the WTA depends on the current level present in the competing input branches. This current level will depend on the values chosen for L_A, L_B, and L_M, as well as on the input pattern **I** and all internal weights $\mathbf{z}_j$. To keep the presentation simple, delay times will be given as a function of T_j values directly. Table 3.2 shows the measured delay times when T_1 changes from T_1^a to T_1^b, and T_2 to T_{18} have the values given in the table. Delay t_{d1} is the time needed by category y_1 to win when T_1 switches from T_1^a to T_1^b, and t_{d2} is the time needed by category y_2 to win when T_1 decreases from T_1^b to T_1^a. As can be seen, this delay is always below $1.2\mu s$.

For the cases when the vigilance criterion is not directly satisfied and hence comparators CC_j cut some of the T_j currents, an additional delay is observed. This extra delay has been measured to be less than $400ns$ for the worst cases. Therefore, the time needed until the WTA selects the maximum among all $\overline{c_j}T_j$ is less than $1.2\mu s + 0.4\mu s = 1.6\mu s$.

Learning Time. After a delay of $1.6\mu s$ (so that the WTA can settle), the learn signal $\overline{LEARN}$ (see Fig. 3.1) is enabled during a time t_{LEARN}. To measure the minimum t_{LEARN} time required, this time was set to a specific value during a training/learning trial, and it was checked that the weights had been updated properly. By progressively decreasing t_{LEARN} until some of the weights did not update correctly, it was found that the minimum time for proper operation was $190ns$. By setting t_{LEARN} to $200ns$ and allowing the WTA a

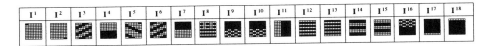

Figure 3.9. Set of Input Patterns

delay of $1.6\mu s$, the total throughput time of the ART1 chip is established as $1.8\mu s$.

Comparison with Digital Neural Processors

A digital chip with a feedforward speed of a connections per second, a learning speed of b connection updates per second, and a WTA section with a delay of c seconds must satisfy the following equation to achieve a throughput time of $1.8\mu s$ when emulating the ART1 algorithm of Fig. 3.1(c):

$$\frac{N \times (2M+1)}{a} + \frac{N}{b} + c = 1.8\mu s \qquad (3.9)$$

Note that there are N synapse weights z_{ij} to update for each pattern presentation, and $N(2M+1)$ feed-forward connections: $N \times M$ connections to generate all terms $T_j = L_A|\mathbf{I} \cap \mathbf{z}_j| - L_B|\mathbf{z}_j| + L_M$, $N \times M$ connections to generate terms $L_A|\mathbf{I} \cap \mathbf{z}_j|$, and N connections to generate $L_A|\mathbf{I}|$.

Assuming $c = 100ns$, and $a = b$, eq. (3.9) results in a processing speed of $a = b = 2.2 \times 10^9$ connections/s or connection-updates/s for $N = 100$ and $M = 18$. A digital neural processor would require such figures of merit to equal the processing time of the analog ART1 chip presented here. Therefore, this "approximate reasoning" makes us conclude that this chip has an equivalent computing power of $a + b = 4.4 \times 10^9$ connections/s plus connection-updates/s.

System Level Performance

Although the internal processing of the chip is analog in nature, its inputs (I_i) and outputs $(\overline{y_j})$ are binary valued. Therefore, the system level behavior of the chip can be tested using conventional digital test equipment. In our case we used the HP82000 IC Evaluation System.

An arbitrary set of 100-bit input patterns $\mathbf{I}^k$ was chosen, shown in Fig. 3.9. A typical clustering sequence is shown in Fig. 3.10, for $\rho = 0.7$ and $\alpha = L_A/L_B = 1.05$. The first column indicates the input pattern $\mathbf{I}^k$ that is fed to the $F1$ layer. The other 18 squares (10×10 pixels) in each row represent each of the internal $\mathbf{z}_j$ vectors after learning is finished. The vertical bars

72 ADAPTIVE RESONANCE THEORY MICROCHIPS

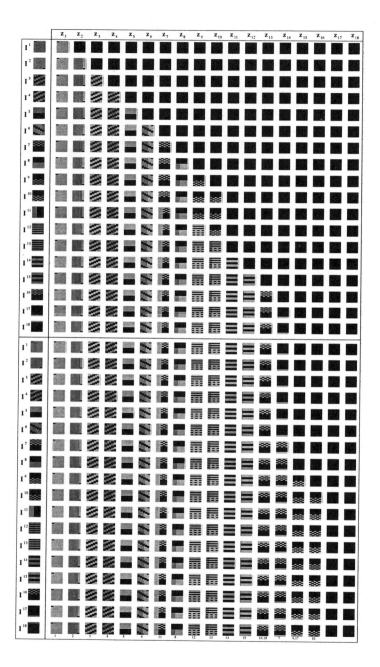

Figure 3.10. Clustering Sequence for $\rho = 0.7$ and $\alpha = L_A/L_B = 1.05$

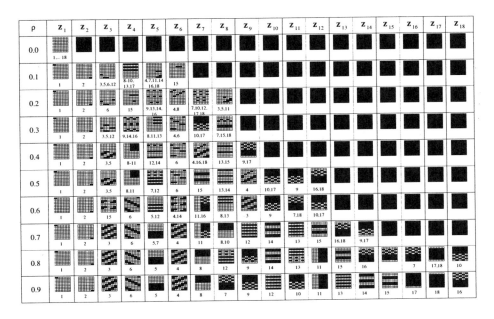

Figure 3.11. Categorization of the Input Patterns for $\alpha = 1.07$ and different values for ρ

to the right of some z_j squares indicate that these categories won the WTA competition while satisfying the vigilance criterion. Therefore, such categories correspond to z_J, and these are the only ones that are updated for that input pattern I^k presentation. The figure shows only two iterations of input patterns presentation, because no change in weights were observed after these. The last row of weights z_j indicates the resulting categorization of the input patterns. The numbers below each category indicate the input patterns that have been clustered into this category. In the following figures we will show only this last row of learned patterns together with the pattern numbers that have been clustered into each category.

Fig. 3.11 shows the categorizations that result when tuning the vigilance parameter ρ to different values while the currents were set to $L_A = 3.2\mu A$, $L_B = 3.0\mu A$, and $L_M = 400\mu A$ ($\alpha = L_A/L_B = 1.07$). Note that below some categories there is no number. This is a known ART1 behavior: during the clustering process some categories might be created which will later be abandoned and will not represent any of the training patterns. In Fig. 3.12 the vigilance parameter is maintained constant at $\rho = 0$, while α changes from 1.07 to 50. For a more detailed explanation on how and why the clustering behavior depends on ρ and α, refer to Chapter 2.

74 ADAPTIVE RESONANCE THEORY MICROCHIPS

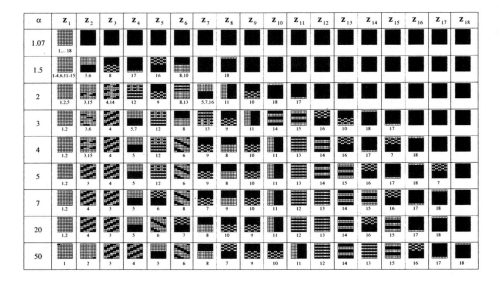

Figure 3.12. Categorization of the Input Patterns for $\rho = 0$ and different values of α

Yield and Fault Tolerance

A total of 30 chips (numbered 1 through 30 in Table 3.3 and Fig. 3.13) were fabricated. For each chip every subcircuit was independently tested and its proper operation verified; 14 different faults were identified. Table 3.3 indicates the faults detected for each of the 30 chips. The faults have been denoted from **F1** to **F14**, and are separated into two groups:

- *Catastrophic Faults* are those clearly originated by a short or open circuit failure. These faults are **F1**, ... **F8**. This kind of faults would produce a failure in a digital circuit.

- *Non-Catastrophic Faults* are those that produce a large deviation from the nominal behavior, too large to be explained by random process parameter variations. These faults are **F9**, ... **F14**. This kind of faults would probably not produce a catastrophic failure in a digital circuit, but be responsible for significant delay times degradations.

Table 3.4 describes the subcircuits where the faults of Table 3.3 were found. Note that the most frequent faults are **F3/F9** and **F4/F10**, which are failures in some current sources L_A or L_B, and these current sources occupy a significant percentage of the total die area. Fault **F1** is a fault in the shift register that loads the input vector $\mathbf{I}^k$. Fault **F2** is a fault in the WTA circuit. Therefore,

Table 3.3. Classification of Faults detected in the 30 fabricated chips

chip #	Catastrophic Faults								Non-Catastrophic Faults					
	F1	F2	F3	F4	F5	F6	F7	F8	F9	F10	F11	F12	F13	F14
1					X						X			
2		X					X		X	X		X	X	
3									X	X		X		X
4		X							X	X		X		
5	X	X							X					
6			X	X					X	X	X			
7			X						X					
8									X					
9		X		X					X	X				
10			X	X					X	X				
11	X													
12					X					X				
13	X	X	X						X	X				
14		X							X					
15		X	X						X	X				
16			X						X					
17	X		X						X					
18									X	X		X		
19										X				
20		X	X	X					X	X		X		
21	X								X					
22									X			X		
23	X								X					
24				X						X				
25														
26									X					
27	X	X	X						X					
28														
29	X													
30									X	X		X		

chips with an **F1** or **F2** fault could not be tested for system level operation[3]. Faults **F3** and **F9** are faults detected in the same subcircuits of the chip, with

---

[3]Chips 9 and 14 had the WTA partially operative and could be tested for system level behavior

Table 3.4. Description of Faults

F1	non-operative shift register for loading $\mathbf{I}^k$
F2	non-operative WTA circuit
F3/F9	fault in a current source L_A
F4/F10	fault in a current source L_B
F5/F11	fault in $1:\rho$ current mirror
F6/F12	fault in current mirror CMM
F7/F13	fault in current mirrors CMA_j or CMB_j
F8/F14	fault in current mirror CMC

F3 being catastrophic and **F9** non-catastrophic. The same is valid for **F4** and **F10**, **F5** and **F11**, and so on until **F8** and **F14**.

Note that only 2 of the 30 chips (6.7%) are completely fault-free (chips 25 and 28). According to the simplified expression for the yield performance as a function of die area Ω and process defects density ρ_D [Strader, 1989],

$$\text{yield}(\%) = 100 e^{-\rho_D \Omega} \qquad (3.10)$$

this requires a process defect density [4] of $\rho_D = 3.2 cm^{-1}$. On the other hand, ignoring the non-catastrophic faults yields 9 out of 30 chips (30%) (chips without catastrophic faults are 3, 8, 18, 19, 22, 25, 26, 28, and 30). According to eq. (3.10) such a yield would be predicted if the process defect density is $\rho'_D = 1.4 cm^{-1}$.

Even though the yield is quite low, many of the faulty samples were still operative. This is due to the fault tolerant nature of the neural algorithms in general [Chu, 1991], [Neti, 1992],[Kim, 1991], and the ART1 algorithm in particular. There were 16 chips with operative shift register and WTA circuit. We performed system level operation tests on these chips to verify if they would be able to form clusters of the input data, and verified that 12 of these 16 chips were able to do so. Moreover, 6 (among which were the two completely fault-free chips) behaved exactly identically. The resulting clustering behavior of these 12 chips is depicted in Fig. 3.13 for $\rho = 0.5$ and $\alpha = 1.07$.

3.7 EXPERIMENTAL RESULTS OF SECOND PROTOTYPE

A second prototype chip was fabricated in a $1.0\mu m$ CMOS process (ES2 through Eurochip) by implementing the L_A and L_B current sources directly as single

---

[4]The effective die area is $\Omega = (0.92 cm)^2$

ART1 AND ARTMAP VLSI CIRCUIT IMPLEMENTATION 77

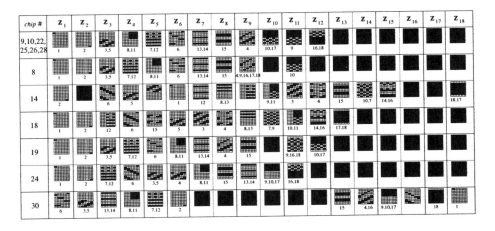

Figure 3.13. Categorization of the Input Patterns performed by Operative Samples

multiple-outputs current mirrors [Serrano-Gotarredona, 1997]. This could be done because at that time we had some information on long distance gradient induced mismatch for this technology. In Appendix C we show how we performed transistor random mismatch characterization using a special purpose chip for this technology. Using the same measurements, we could obtain some information (limited but reasonably valid) about the gradient induced mismatch behavior for this technology. In what follows we show first how we extracted the gradient induced mismatch from the chip described in Appendix C, and afterwards we show characterization results from the second ART1 chip.

Gradient Induced Mismatch Information from Mismatch Characterization Chip

The mismatch characterization chip described in Appendix C, fabricated in a $1.0\mu m$ CMOS process (ES2 through Eurochip), contains an 8×8 array of cells. Each cell contains 30 NMOS and 30 PMOS transistors of different sizes. The size of the complete array is $2.5mm \times 2.5mm$ and distance between equal transistors is about $300\mu m$. Using this characterization chip we found out that for transistors of size $10\mu m \times 10\mu m$ spread over a chip area of $2.5mm \times 2.5mm$, biased by the same gate-to-source V_{GS} and drain-to-source V_{DS} voltages so that their nominal current was around $10\mu A$, the measured currents are as depicted in Fig. 3.14. Fig. 3.14(a) shows, as a function of transistor position, the current measured for each transistor of an NMOS array. Fig. 3.14(b) shows the same for a PMOS array. As can be seen, the surfaces present a long distance gradient component and a short distance noise component. Let us call the measured

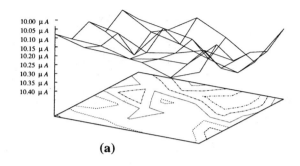

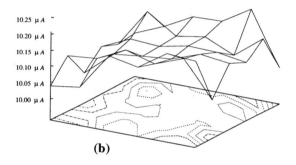

Figure 3.14. Measured Current for an Array of MOS Transistors with the same V_{GS} and V_{DS} voltages (for a nominal current of $10\mu A$), spread over a die area of $2.5mm \times 2.5mm$. (a) Array of NMOS Transistors, (b) Array of PMOS Transistors.

currents surface $I_o(x,y)$. For this surface we can compute the best fit plane $I_o^p(x,y) = Ax + By + C$. Then, for each point (x,y) we can define

$$\Delta I_o(x,y) = I_o(x,y) - I_o^p(x,y) \qquad (3.11)$$

By computing the standard deviation of $\Delta I_o(x,y)$, $\sigma(\Delta I_o)$, we are extracting the noise component of surface $I_o(x,y)$. The gradient component is defined by plane $I_o^p(x,y)$. The maximum deviation due to the gradient component is given by

$$\Delta I_o^p = max\{I_o^p(x,y)\} - min\{I_o^p(x,y)\} \qquad (3.12)$$

Table 3.5. Current Mismatch Components for Transistor Arrays with $10\mu A$ nominal current, $10\mu m \times 10\mu m$ Transistor Size, and $2.5mm \times 2.5mm$ die area, for the ES2-1.0μm CMOS process.

chip	NMOS				PMOS			
	$\sigma(\Delta I_o)(\%)$	$\Delta I_o^p(\%)$	r	$\sigma_T(I_o)(\%)$	$\sigma(\Delta I_o)(\%)$	$\Delta I_o^p(\%)$	r	$\sigma_T(I_o)(\%)$
1	0.57	1.30	2.652	0.67	0.58	1.53	2.278	0.67
2	0.62	1.98	1.874	0.83	0.47	0.74	3.830	0.51
3	0.47	3.09	0.921	0.79	0.48	0.82	3.519	0.51
4	0.52	0.90	3.456	0.56	0.40	2.18	1.100	0.63
5	0.54	1.65	1.959	0.64	0.46	0.60	4.666	0.49
6	0.58	3.01	1.160	0.88	0.45	2.18	1.236	0.72
7	0.65	1.96	1.996	0.82	0.44	0.83	3.171	0.50
8	0.73	2.15	2.027	0.90	0.41	1.28	1.926	0.50

On the other hand, for the noise component, 98% of the points remain within the $\pm 3\sigma(\Delta I_o)$ interval. Consequently, let us define the maximum deviation due to the noise component as $6\sigma(\Delta I_o)$. Let us now define

$$r = \frac{6\sigma(\Delta I_o)}{\Delta I_o^p} \qquad (3.13)$$

as the ratio between noise component and gradient component contributions. Table 3.5 shows these ratios measured for NMOS and PMOS transistors of size $10\mu m \times 10\mu m$, driving nominal currents of $10\mu A$ and for different chips. Also shown in Table 3.5 are the standard deviations of the noise component $\sigma(\Delta I_o)$, the maximum deviation of the gradient component ΔI_o^p, and the total standard deviation of transistor currents $\sigma_T(I_o)$, computed as

$$\sigma_T(I_o) = \sqrt{\overline{I_o^2} - \overline{I_o}^2} \qquad (3.14)$$

Details of Second Prototype Results

The current mirror tree-like structure of Fig. 3.6 was intended to suppress the gradient component of a $1cm^2$ chip. The noise component can only be reduced by increasing transistor area [Pelgrom, 1989], [Appendix C]. Table 3.5 reveals that for die areas of $2.5mm \times 2.5mm$, transistor sizes of $10\mu m \times 10\mu m$, and nominal currents of $10\mu A$, the contribution of noise component is equal or higher than the gradient component, while the standard deviation of current mismatch $\sigma_T(I_o)$ is kept below 1%. Consequently, for these dimensions we can

Table 3.6. Measured Current Error for L_{A1} NMOS Current Sources

chip #	$\sigma(\Delta I_o)(\%)$	$\Delta I_o^p(\%)$	r	$\sigma_T(I_o)(\%)$
1	0.63	1.39	2.694	0.71
2	0.51	0.68	5.311	0.62
3	0.68	1.91	2.128	0.81
4	0.59	0.28	12.607	0.61
5	0.64	1.27	3.204	0.69
6	0.65	1.29	3.028	0.71
7	0.66	0.41	9.535	0.68
8	0.64	0.92	4.174	0.67
9	0.79	2.20	2.157	0.91
10	0.74	0.43	10.368	0.75

avoid the use of high area consuming circuit structures (like common centroid mirrors arranged in a tree-like fashion) to eliminate the gradient component, and directly implement a single current mirror with all the outputs needed. This is the approach we used in this second ART1 prototype chip. This chip has a die area of $2.5mm \times 2.2mm$, and contains an array of 50×10 synapses, each synapse with two L_A and one L_B current sources. The current sources transistors are of size $10\mu m \times 10\mu m$ and drive a nominal current of $10\mu A$. Fig. 3.15 shows the measured currents of the arrays. Tables 3.6 and 3.7 show the measured values of the mismatch components of the two L_A current sources arrays and Table 3.8 for the L_B current sources array, for all fabricated chips. Note that the total current mismatch standard deviation $\sigma_T(I_o)$ is less than 1% for all chips.

Due to the much smaller chip area its fabrication cost is much less than for the first ART1 prototype and its yield performance is significantly higher: 98% by applying eq. (3.10) [5]. This design implements an ART1 system with $N = 50$ input nodes and $M = 10$ category nodes, and the arrays of current sources L_A and L_B were realized directly as current mirrors of 1000 and 500 outputs, respectively, spread over an area of $2.75mm^2$. All ten fabricated chip samples were fully operational and for none of them we were able to detect any fault in its subcircuits. All system components could be isolated and independently characterized. The circuit performances of the different subcircuits were similar to those of the first prototype, and consequently their characteristics will not be repeated. Here we will only provide some illustrative examples on system level behavior.

---

[5] All 10 fabricated chips were completely fault free

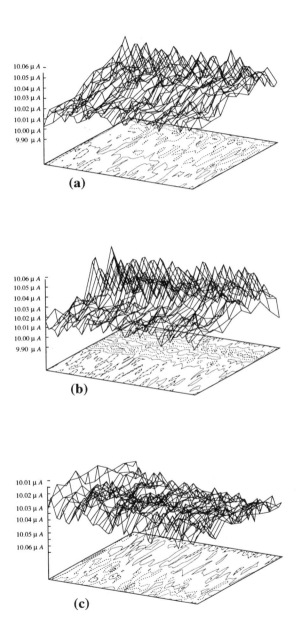

Figure 3.15. Measured Synaptic Current as a Function of Synapse Position in the Chip, for (a) L_{A1} Current Sources, (b) L_{A2} Current Sources, and (c) L_B Current Sources.

Table 3.7. Measured Current Error for L_{A2} NMOS Current Sources

chip #	$\sigma(\Delta I_o)(\%)$	$\Delta I_o^p(\%)$	r	$\sigma_T(I_o)(\%)$
1	0.69	1.36	3.057	0.75
2	0.64	0.60	6.391	0.65
3	0.68	1.93	2.097	0.80
4	0.68	0.15	27.027	0.70
5	0.64	1.27	3.050	0.71
6	0.71	1.26	3.384	0.78
7	0.65	0.40	9.818	0.66
8	0.63	1.24	3.047	0.69
9	1.08	1.30	4.962	1.12
10	0.88	0.54	9.776	0.90

Table 3.8. Measured Current Error for L_B PMOS Current Sources

chip #	$\sigma(\Delta I_o)(\%)$	$\Delta I_o^p(\%)$	r	$\sigma_T(I_o)(\%)$
1	0.62	0.62	6.076	0.64
2	0.59	0.22	16.497	0.60
3	0.56	3.32	1.015	0.89
4	0.63	0.90	4.196	0.64
5	0.65	1.83	2.118	0.76
6	0.64	1.49	2.565	0.73
7	0.60	1.58	2.255	0.67
8	0.62	1.48	2.524	0.71
9	0.63	0.37	10.080	0.63
10	0.57	2.16	1.573	0.73

ART1 Systems

To test the system behavior, the chip was trained with a set of ten $7 \times 7 = 49$bit input patterns. Each pattern represents each of the ten digits from '0' to '9'. The last input pixel was always set to zero and it is not shown in the figures. The clustering test was repeated for different values of the vigilance parameter ρ and several values of parameter $\alpha = L_A/L_B$.

Fig. 3.16 shows the training sequence for $\rho = 0.3$ and $\alpha = 1.1$. The first column represents the input pattern applied to the system. The remaining ten columns correspond to the weights $\mathbf{z}_j$ stored in each category when the input pattern has been classified and learned. For each line the category with a bar to

ART1 AND ARTMAP VLSI CIRCUIT IMPLEMENTATION   83

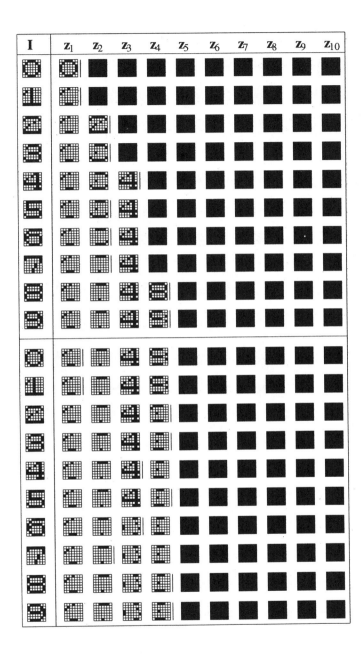

Figure 3.16. Training Sequence for a one-chip ART1 System with $\rho = 0.3$ and $\alpha = 1.1$

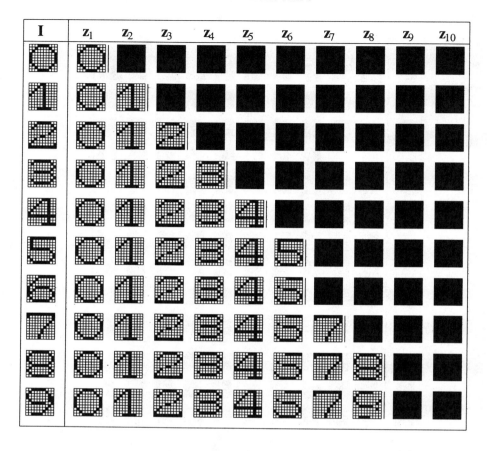

Figure 3.17. Training Sequence for a two-chip ART1 System with $\rho = 0.5$ and $\alpha = 2$

its right is the winning category after the Winner-Takes-All competition. In this case, learning self-stabilizes after two sequences of input patterns presentations. That is, no modification of the winning categories or the stored weights take place in subsequent presentations of the input pattern sequence. As shown in Fig. 3.16, the system has clustered all ten input patterns into four categories.

A two-chip ART1 system was assembled by using the horizontal expansion approach discussed in Section 3.4. In this case, the input patterns had $N = 100$ binary pixels. Fig. 3.17 depicts a training sequence performed on this system. The system classifies the 10 input patterns into 8 categories after a single presentation of the input pattern set. The sequence of Fig. 3.17 was

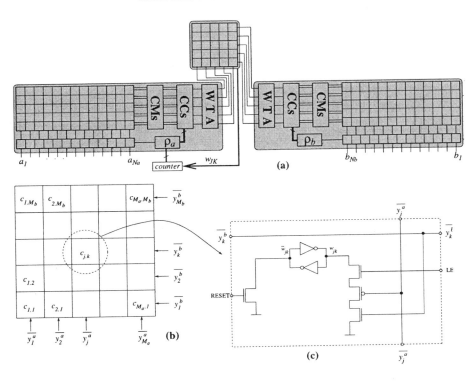

Figure 3.18. (a) ARTMAP Hardware Assembly, (b) Diagram of Inter-ART chip, (c) Detail of Inter-ART Chip Cell

obtained for a vigilance parameter of $\rho = 0.5$, and $\alpha = 2$ ($L_A = 10\mu A$, $L_B = 5\mu A$).

ARTMAP System

An ARTMAP hardware system can be assembled using two ART1 chips and an extra chip for the Inter-ART module, as is shown in Fig. 3.18(a). This architecture corresponds to the "General ARTMAP" architecture discussed in Section 1.3 and Fig. 1.9. The Inter-ART chip, shown in Fig. 3.18(b), is simply an array of cells c_{jk} whose simplified schematic is depicted in Fig. 3.18(c). Each cell has a latch which is set initially to '1' and changes to '0' if $y_j^a = 1$, $y_k^b = 0$, and the $LEARN$ signal is high. Extra transistors, not shown in Fig. 3.18(c), are also included to read out the weight values. During training mode the value of weight w_{JK} is used to control a digital counter that increments the value of ρ_a. If $w_{JK} = 0$ the counter will increase its value until the ART1a winning category changes and w_{JK} becomes '1'. At this moment the counter stops and its content represents the appropriate value for ρ_a.

The system level operation of the ARTMAP hardware system has also been tested using a digital test equipment. Fig. 3.19 shows a system training sequence. The first column, named **a**, represents the input patterns applied to the ART1a chip. The column named **b** represents the input patterns applied to the ART1b chip. The columns named $\mathbf{z}_j^a$ and $\mathbf{z}_k^b$ represent the stored weights in the ART1a and ART1b categories after the classification and learning of each input pattern pair. The categories with a bar to their right are the ones that remain active after the search process has finished, and these are the only ones that are updated with learning. Below each ART1a winning category the final value of the vigilance parameter ρ_a needed in the search process to choose this category is indicated (ρ_a was increased in steps of $\Delta\rho_a = 1/32$). The last column shows the stored weights in the inter-ART module which represent the learned correspondence between the ART1a and ART1b categories. The vigilance parameter ρ_a was initially set to '0' and the current ratio parameter was $\alpha^a = \alpha^b = 2$ ($L_A^{a,b} = 10\mu A$ and $L_B^{a,b} = 5\mu A$). For the ART1b system it was $\rho_b = 0.75$. For this vigilance parameter, the ART1b chip forms a different category for each input pattern.

During the prediction sequence as depicted in Fig. 3.20, only an input pattern **a** is applied to the system. The first column shows the input patterns applied to the system. These input patterns are the result of applying random noise to the training patterns. The ART1a category with a bar to its right is the one chosen by this chip after the search process and the ART1b category with a bar to its right is the one the ART1a category has learned to predict through the Inter-ART weights.

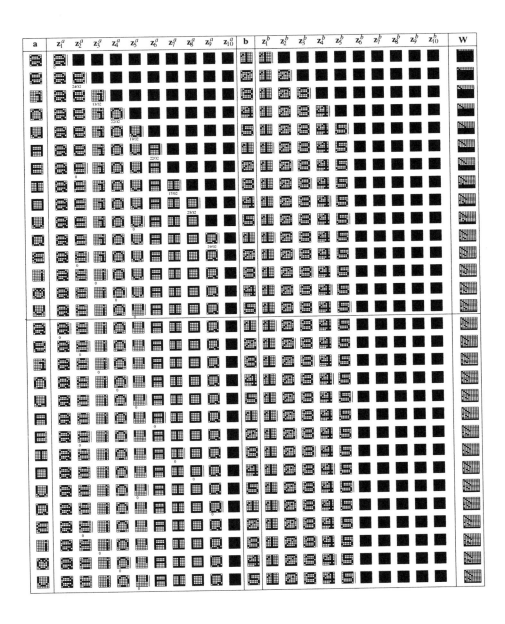

Figure 3.19. Complete Training Sequence of the ARTMAP System for $\overline{\rho_a} = 0$ and $\rho_b = 0.75$

Figure 3.20. Recognition Sequence performed on a previously trained ARTMAP System. Applied Input Patterns are Noisy Versions of the Training Set.

4 A HIGH-PRECISION CURRENT-MODE WTA-MAX CIRCUIT WITH MULTI-CHIP CAPABILITY

4.1 INTRODUCTION

Winner-Takes-All (or Looser-Takes-All) and MAX (or MIN) circuits are often fundamental building blocks in neural and/or fuzzy hardware systems [Haykin, 1994],[Sánchez-Sinencio, 1992],and [Bezdec, 1992]. In Chapter 3 we saw that sometimes one might want to distribute a complex system that includes a Winner-Takes-All (WTA) circuit among several chips. Standard WTAs reported in literature suffer from severe matching degradations when split into several chips. In this Chapter we describe a novel circuit that does not have this drawback.

The operation of WTA-MAX or LTA-MIN circuits is as follows: given a set of M external inputs $(T_1, T_2, \ldots, T_j, \ldots, T_M)$, their operation consists in determining which input j presents the largest (or smallest) value, or what is this maximum (or minimum) value, respectively. If a Winner-Takes-All (WTA) or MAX circuit is available, a Looser-Takes-All (LTA) or MIN circuit is obtained by simply inverting the input $(-T_1, -T_2, \ldots, -T_j, \ldots, -T_M)$[1]. Hence, this Chapter will only concentrate on WTA and MAX circuits.

---
[1]Optionally, a common offset term may be added

In literature, the physical implementation of these systems has been tackled through two main approaches:

1. Systems of $O(M^2)$ complexity: their connectivity increases quadratically with the number of input variables [Elias, 1975], [Kohonen, 1989], [Yuille, 1995], [He, 1993], [Linares-Barranco, 1992], and

2. Systems of $O(M)$ complexity: their connectivity increases linearly with the number of inputs [Lazzaro, 1989], [Choi, 1993].

In a system of $O(M^2)$ complexity, as shown in Fig. 4.1(a), there is one cell per input; each cell has an inhibitory connection (black triangle) to the rest of the cells and an excitatory connection (white triangle) to itself. Therefore, the system has M^2 connections. Each cell j receives an external input T_j. The cell that receives the maximum input will turn all other cells OFF and will remain ON. If the system is a Winner-Takes-All (WTA) circuit, each cell has a binary output that indicates whether the cell is ON or OFF. In a MAX circuit the winning cell will copy its input to a common output.

Under some circumstances[2] it is possible to convert the $O(M^2)$ topology of Fig. 4.1(a) into an $O(M)$ one, as shown in Fig. 4.1(b). In these cases, a global inhibition term is computed. Each cell contributes to this global inhibition, and each cell receives the same global inhibition. Note that now, each cell contributes to inhibit itself. Consequently, the excitatory connection that each cell has to itself must be increased to compensate for this fact.

Typical $O(M)$ WTA circuits reported in literature [Lazzaro, 1989], [Choi, 1993] correspond to the topology shown in Fig. 4.1(c). In such circuits there are also M cells, each receiving an external input T_j. Each cell connects to a common node, through which a global property (for example, a current) is shared between all cells. The amount of that global property taken by each cell depends (nonlinearly) on how much its input T_j deviates from an 'average' of all inputs. Usually this 'average' is not an exact linear average, but is somehow nonlinearly dependent on all inputs. The cell with the maximum input T_j takes most (or all) of the common global property leaving the rest with little or nothing. Due to the way this global property is shared and how the 'average' is computed, the operation of these circuits relies on the matching of transistor threshold voltages of an array of transistors [Lazzaro, 1989], or other transistor parameters. The number of transistors in the array equals, at least, the number of inputs M of the system. If the WTA or MAX circuit has such a large number of inputs so that it must be distributed among different chips, the matching of threshold voltages (or other transistor parameters) will degrade significantly, and the overall system will loose precision in its operation.

---

[2]If the inhibition that goes from cell i to cell j does not depend on j.

A HIGH-PRECISION CURRENT-MODE MULTI-CHIP WTA-MAX CIRCUIT

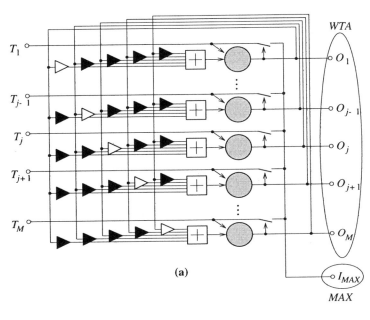

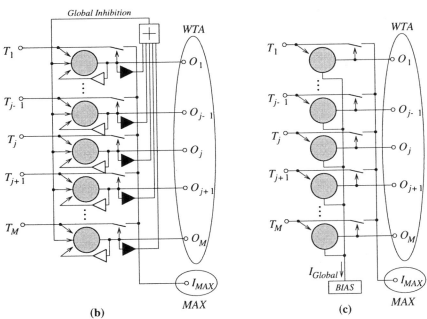

Figure 4.1. WTA topologies. (a) WTA of $O(M^2)$ complexity, (b) transformation to $O(M)$ complexity, (c) typical topology of $O(M)$ WTA hardware implementation

92   ADAPTIVE RESONANCE THEORY MICROCHIPS

This Chapter describes an $O(M)$ complexity circuit technique (which can be represented by the topology in Fig. 4.1(b)) for implementing either WTA and/or MAX circuits, based on current-mode principles. The resulting circuit does not rely on the matching of an M-size transistor array. The precision of the overall system relies on precise current replication, which can be achieved locally without matching M transistors. Sometimes, when assembling large neural and/or fuzzy systems, a WTA/MAX circuit must be distributed among several chips [Serrano-Gotarredona, 1997]. The circuit described here can be distributed among several chips with no influence on its precision, as shown in the Section on experimental results.

In the next Section a mathematical model that performs WTA/MAX operation is described. This operation principle will be used in Section 4.3 to develop a current-mode processing circuit. Sections 4.4 and 4.5 deal with stability considerations of the circuit presented in Section 4.3. Finally, Section 4.6 provides experimental measurement results obtained from prototypes fabricated in two different CMOS technologies, and from two-chip systems formed by chips of the same or different technologies.

4.2   OPERATION PRINCIPLE

The operation principle given in this Section can be used for simultaneous implementation of a WTA and MAX circuit. The system has M cells. Each cell j produces an output

$$I_{oj} = \alpha_j H(T_j - I_o) \quad , \quad j = 1, \ldots, M \tag{4.1}$$

where

$$I_o = \sum_{j=1}^{M} I_{oj}, \tag{4.2}$$

$H(\cdot)$ is the step function defined as

$$H(x) = \begin{cases} 1 & , \quad x > 0 \\ \in [0,1] & , \quad x = 0 \\ 0 & , \quad x < 0 \end{cases} \tag{4.3}$$

and T_j is the external input to the j-th cell. Substituting eq. (4.1) into eq. (4.2) yields

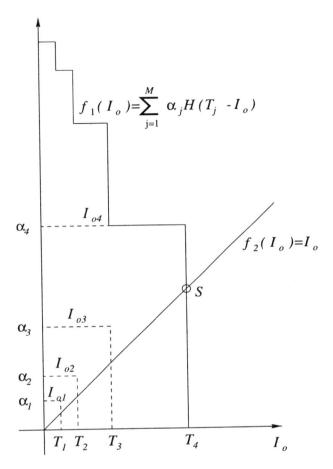

Figure 4.2. Graphic Representation of the Solution of eq. (4.4)

$$I_o = \sum_{j=1}^{M} \alpha_j H(T_j - I_o) \quad (4.4)$$

Fig. 4.2 shows a graphic representation of the functions $f_1(I_o) = \sum_j \alpha_j H(T_j - I_o)$ and $f_2(I_o) = I_o$. The intersection of $f_1(I_o)$ and $f_2(I_o)$ provides the solution to eq. (4.4). Note that if $\alpha_j > 0 \ \forall j$, eq. (4.4) has a unique equilibrium point S, as deduced from Fig. 4.2. Furthermore, if

$$\alpha_j \geq T_j \quad , \quad \forall j \quad (4.5)$$

the value of I_o at the equilibrium point S is

$$I_o|_S = max\{T_j\} \qquad (4.6)$$

and the cell that drives a nonzero output $I_{oj} \neq 0$ is the winner. Consequently, a circuit that implements eq. (4.4) can be used to realize both a WTA or a MAX circuit.

In the case of an LTA or a MIN circuit, the same mathematical model of Fig. 4.2 applies if each input equals

$$I_L - T_j, \qquad (4.7)$$

where I_L is an upper bound for all inputs

$$0 \leq T_j \leq I_L \quad , \quad \forall j \qquad (4.8)$$

4.3 CIRCUIT IMPLEMENTATION

This Section shows how to realize a circuit that implements eq. (4.4) using currents to represent the mathematical variables T_j and I_o. The circuit for each cell j is shown in Fig. 4.3. It consists of a 2-output current mirror, a MOS transistor, and a digital inverter. Each cell j receives two input currents, T_j and I_o, and delivers one output current I_{oj}. The inverter acts as a current comparator. If $I_o > T_j$ the inverter output v_{oj} is low, the MOS transistor is OFF, and I_{oj} is zero. If $I_o < T_j$ the inverter output is high, the MOS transistor is ON, and $I_{oj} = T_j$. Consequently, the circuit of Fig. 4.3 implements a cell with $\alpha_j \equiv T_j$.

Fig. 4.4 shows the complete WTA or MAX circuit. It consists of M cells shown in Fig. 4.3 and an additional M-output current mirror. Note that the responsibility of the M-output current mirror is to deliver the sum of currents $I_o = \sum_j I_{oj}$ to each of the M cells. Replication and transportation of current I_o must be very precise. If the number of cells M is too large, or if the circuit has to be distributed among several chips, high precision in I_o replication cannot be guaranteed by a single current mirror with M outputs. In this case, replication of current I_o must rely on several mirrors with a smaller number of outputs each but with guaranteed precise replication. Fig. 4.5 shows an arrangement to distribute the circuit of Fig. 4.4 among several chips. The fact that current I_o can be replicated many times without relying on the matching of a large array of transistors is the advantage of this WTA and MAX (or LTA and MIN) circuit technique over others.

A HIGH-PRECISION CURRENT-MODE MULTI-CHIP WTA-MAX CIRCUIT

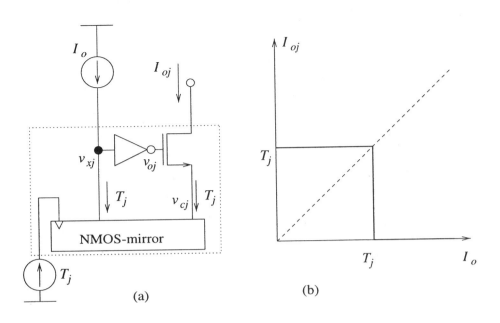

Figure 4.3. WTA unit cell: (a) circuit diagram, (b) transfer curve

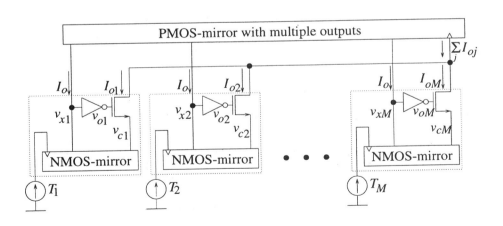

Figure 4.4. Diagram of the WTA circuit

The precision of the overall WTA current mode circuit is determined by the kind of current mirrors used. Since the current comparator has virtually no offset, the current error at the input of each current comparator is determined

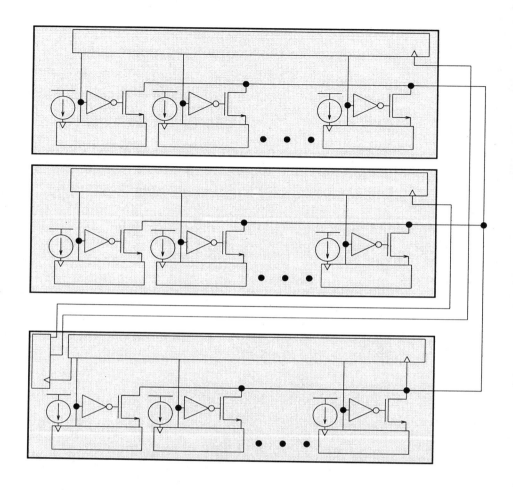

Figure 4.5. Strategy to assemble several chips

by mirrors induced mismatches. The standard deviation of the error at the positive current $\sigma(I_o)$ available at the input of each current comparator results from one p-mirror reflection, preceded by an n-mirror reflection of the winning cell. If the error introduced by the p-mirror can be statistically characterized by standard deviation σ_P, and that of the n-mirror by σ_N

$$\sigma^2(I_o) = \sigma_N^2 + \sigma_P^2 \tag{4.9}$$

The error of the negative current T_j at the current comparator inputs results from a single n-mirror reflection,

$$\sigma^2(T_j) = \sigma_N^2 \tag{4.10}$$

The total current error at the input of each current comparator is therefore given by,

$$\sigma_{Total}^2 = \sigma^2(I_o) + \sigma^2(T_j) = 2\sigma_N^2 + \sigma_P^2 \tag{4.11}$$

However, the error introduced by a current mirror is not only the random mismatch contribution, as considered in eqs. (4.9)-(4.11), but also its systematic error contribution, which results from different drain-to-source voltages at the reflecting transistors and poor impedance coupling. Fig. 4.6 shows several current mirror topologies which intend to improve precision performance with respect to the simple current mirror.

The active-input current mirror idea (see Fig. 4.6(a)) [Nairn, 1990a] was introduced as a need to maintain a constant voltage at the input of a current mirror in order to avoid current subtraction errors in previous stages. This technique allows the V_{GS} voltage of the current mirror input transistor $M1$ to be independent of its V_{DS} voltage, which will be kept constant, and therefore lowers the mirror input impedance which minimizes loading effects on previous stages. The current mirror will be operative as long as transistors $M1$ and $M2$ are kept in saturation. The higher the reference voltage V_D is, the more current is allowed through the mirror with $M1$ operating in saturation. The drain-to-source voltage of the output transistor depends on the load of the mirror and will be dependent on the mirror current. If the load impedance is not sufficiently low $M2$ will suffer of large drain-to-source voltage variations, which through the channel length modulation effect, will cause a systematic mismatch error between the input and output currents of the mirror. Such a circumstance can be avoided by using the cascode current mirror of Fig. 4.6(b). However, this mirror has a smaller output voltage swing and requires a high input voltage drop for high currents (which are needed for maximum accuracy) [3.16]. The regulated-cascode output stage (see Fig. 4.6(c)[3]) [Sackinger, 1990] would allow to maintain the V_{DS} of transistor $M2$ constant and to increase significantly the output impedance of the mirror, while not sacrificing voltage range at the input of the current mirror. This current mirror will be operative as long as transistor $M2$ remains in saturation, which depends on V_D and the input current, as well as on the load impedance at the output of the mirror.

By combining the active current mirror input of Fig. 4.6(a) with the regulated-cascode output in Fig. 4.6(c), the *active-input regulated-cascode current mirror*

---

[3]Although we are showing a differential input voltage amplifier, in the original paper [Sackinger, 1990] a single input amplifier was used. In this case the voltage V_D would be set through process and circuit parameters.

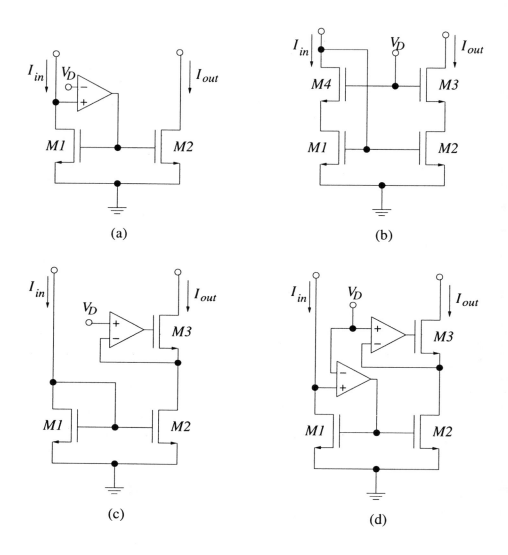

Figure 4.6. Enhanced current mirror topologies, (a) active, (b) cascode, (c) regulated cascode output, and (d) active regulated cascode

of Fig. 4.6(d) results [Serrano-Gotarredona, 1994]. This current mirror has a very low input impedance, a very high output impedance and is operative if transistors $M1$ and $M2$ are either in saturation or ohmic region (because their V_{DS} voltages are always equal). Therefore, the gate voltage of transistors $M1$ and $M2$ can change almost from rail to rail. However, this current mirror will fail to operate with high accuracy if the output voltage approaches V_D. But, on the other hand, V_D can be made smaller than for Fig. 4.6(a) because $M1$ and $M2$ can operate now in ohmic regime.

When cascading current mirrors, the regulated-cascode output is not needed and the configuration of Fig. 4.6(a) can be used. The virtual ground effect at the drain of transistor $M2$ is produced by the active input of the next current mirror. However, in this case V_D has to be set equal for all PMOS and NMOS active current mirrors. Also, care has to be taken by choosing the value of V_D in order to respect the output voltage range of the circuit at the input of the first mirror, and to respect the input voltage range of the circuit at the output of the last current mirror.

As has been previously explained, the accuracy limitation of current mirrors has two main types of sources, systematic errors and random errors. Systematic errors are caused by different V_{DS} voltages at transistors $M1$ and $M2$ due to poor input/output impedance coupling between subsequent stages and high-order nonlinear effects. Random errors are fundamentally caused by differences in the electrical parameters between transistors $M1$ and $M2$, due to random process parameter variations. While random mismatch error contributions are practically independent on the circuit topology, systematic errors change considerably from one topology to another. We will consider that the total precision of a current mirror is given by

$$\Delta I_{Total} = \Delta I_{sys} + \sigma_I \qquad (4.12)$$

where ΔI_{sys} is the systematic error contribution (which can be evaluated through a single nominal Hspice simulation) and σ_I is the standard deviation of the output current (which can be evaluated through 30 Hspice Monte Carlo simulations). The statistical significance of σ_I is that 68% of the samples have an output current error within the range $(\Delta I_{sys} - \sigma_I, \Delta I_{sys} + \sigma_I)$. Fig. 4.7 shows a comparison of the precision performance of the topologies in Fig. 4.6, based on electrical simulations. For random mismatch errors Hspice simulations, it is considered that the only sources of random mismatch are the differences in threshold voltage (V_T) and current factor ($\beta = C_{ox}\mu W/L$), and that their standard deviation is given by [Pelgrom, 1989],

$$\sigma^2(V_T) = \frac{A_{V_T}^2}{WL} + S_{V_T}^2 D^2 \qquad (4.13)$$

$$\frac{\sigma^2(\beta)}{\beta^2} = \frac{A_\beta^2}{WL} + S_\beta^2 D^2$$

where W and L are the sizes of the transistors, $D(\sim W)$ their separation, and $A_{V_T} = 15mV\mu m$, $S_{V_T} = 2\mu V/\mu m$, $A_\beta = 2.3\%\mu m$, $S_\beta = 2 \times 10^{-6}\mu m$

(parameters given in [Pelgrom, 1989] for a $1.6\mu m$ N-well process with $25nm$ gate oxide and direct wafer writing).

In the following simulations we use $W = 100\mu m$ and $L = 20\mu m$ for transistors $M1$ and $M2$ of all current mirror topologies, with $V_D = 1.5V$ (power supply is $5V$). Fig. 4.7 represents the total accuracy of eq. (4.12) as a function of operating current for different current mirror topologies [Serrano-Gotarredona, 1994]. The first two topologies are for the *active-input regulated-cascode current mirror* with amplifier gain values $A = 1000$ and $A = 100$. For $A = 1000$ the resolution is above 8-bit for current values between $40\mu A$ and $750\mu A$, while for $A = 100$ it is only between $40\mu A$ and $330\mu A$. In both cases the decrease in precision above $\sim 300\mu A$ is because transistors $M1$ and $M2$ enter their ohmic region of operation. A simple current mirror has a resolution below 8-bit for the complete current range. This is mainly due to poor impedance coupling, which can be avoided with regulated cascode outputs, achieving 8-bit resolution between $45\mu A$ and $150\mu A$. The cascode mirror suffers from large voltage drops at its input, thus offering the 8-bit resolution only between $40\mu A$ and $70\mu A$. The active-input mirror (Fig. 4.6(a)) has low resolution because its output voltage was set to $2.5V$, hence rendering an important systematic error contribution. Also shown in Fig. 4.7 is the random mismatch error contribution (σ_I in eq. 4.12), which is approximately the same for all topologies. Note that all topologies suffer from loss of precision at high currents. This is produced because the increasing voltage drops at the different nodes approximate the limit of available voltage range.

Table 4.1 shows the transistor level simulated transient times of our design when using a regular OTA [Degrauwe, 1982] as the active device. These transients correspond to the time the mirror takes to settle to 1% of the final current value, when the input is driven by an ideal current step signal changing between the levels given in the first row of the table. Table 4.1 also shows the delays for other topologies that have the same transistor sizes. Note that speed is improved with respect to mirrors that do not use active devices. The reason is that the active device introduces more design variables, thus allowing a more optimum final result.

As we will see in Section 4.6, we built a CMOS prototype of the circuit in Fig. 4.4 (and Fig. 4.5) using simple current mirrors for the NMOS mirrors, and active-input mirrors for the PMOS and the assembling current mirrors. As it will be discussed in Section 4.6, this is sufficient to guarantee multichip WTA operation. For precise MAX operation, however, more elaborated mirrors would be needed for all mirrors.

4.4 SYSTEM STABILITY COARSE ANALYSIS

Let us assume that the dynamics of each cell (see Fig. 4.3) can be modelled by the following first-order nonlinear differential equation

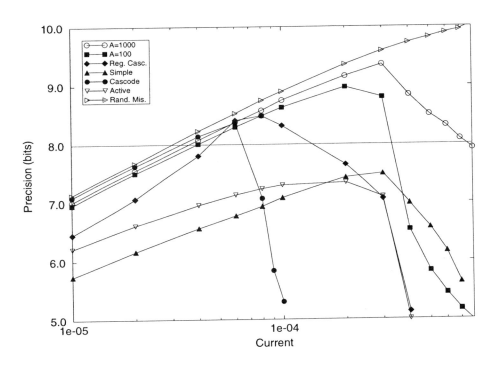

Figure 4.7. Resolution (in bits) as a function of working current for different current mirror topologies

Table 4.1. Simulated Transient Times

	$10\mu A$ to $15\mu A$	$50\mu A$ to $75\mu A$	$100\mu A$ to $150\mu A$	$250\mu A$ to $375\mu A$	$400\mu A$ to $600\mu A$	$530\mu A$ to $800\mu A$
act. reg. casc.	65ns	70ns	45ns	45ns	70ns	460ns
simple	160ns	80ns	65ns	40ns	80ns	35ns
cascode	180ns	85ns	-	-	-	-
active input	65ns	70ns	45ns	35ns	45ns	-
reg. cas. out.	165ns	80ns	65ns	50ns	-	-

$$C_c \dot{v}_{xj}(t) + G_c(v_{xj}(t) - v_M) + T_j = I_o(t) \tag{4.14}$$

where C_c is the total capacitance available at node v_{xj}, G_c is the total conductance at this node, and v_M is the inverter trip voltage. Let us also assume that the output current of a cell is given by

$$I_{oj}(t) = T_j U(v_M - v_{xj}(t)) \tag{4.15}$$

where $U(\cdot)$ is a continuous and differentiable approximation to the step function of eq. (4.3). For example, we can define $U(\cdot)$ as the following sigmoidal function,

$$U(x) = \frac{1}{1 + e^{-x/\epsilon}} \tag{4.16}$$

where ϵ is positive and non-zero but close to zero. Now consider eq. (4.14) for two nodes, j and w. Let w be the node that eventually should become the winner. If we subtract eq. (4.14) for the two nodes j and w, then

$$C_c[\dot{v}_{xj}(t) - \dot{v}_{xw}(t)] + G_c[v_{xj}(t) - v_{xw}(t)] = T_w - T_j \tag{4.17}$$

Eq. (4.17) has the following solution

$$v_{xj}(t) - v_{xw}(t) = \frac{T_w - T_j}{G_c} + \left[v_{xj}(0) - v_{xw}(0) - \frac{T_w - T_j}{G_c}\right] e^{-\frac{t}{\tau_c}}$$
$$\tau_c = \frac{C_c}{G_c} \tag{4.18}$$

After a few time constants τ_c the difference between the two node voltages will remain constant and equal to their difference at the equilibrium point. Therefore, if we can obtain the expression for $v_{xw}(t)$, applying eq. (4.17) would obtain $v_{xj}(t)$ for the rest of the nodes.

Consider now eq. (4.14) for node w, and substitute eqs. (4.2) and (4.15) into it,

$$C_c \dot{v}_{xw}(t) + G_c(v_{xw}(t) - v_M) + T_w = \sum_j T_j U(v_M - v_{xj}(t)) \tag{4.19}$$

Since $v_{xj}(t)$ is given by eq. (4.17), after a few time constants τ_c eq. (4.19) becomes

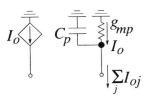

Figure 4.8. Small Signal Modeling of Delay of the N-output Current Mirror

$$C_c \dot{v}_{xw}(t) = G_c(v_M - v_{xw}(t)) - T_w + \\ + \sum_j T_j U\left(v_M - v_{xw}(t) - \frac{T_w - T_j}{G_c}\right) \quad (4.20)$$

This first order differential equation has stable equilibrium points if

$$\left.\frac{d\dot{v}_{xw}}{dv_{xw}}\right|_{\text{equilibrium point}} < 0 \quad (4.21)$$

By deriving eq. (4.20) with respect to v_{xw} results

$$C_c \frac{d\dot{v}_{xw}}{dv_{xw}} = -G_c - \sum_j T_j U'(\cdot) \quad (4.22)$$

Since G_c, T_j, and $U'(\cdot)$ are always positive, eq. (4.22) is always negative for all possible values of v_{xw} (including the equilibrium point). Consequently, eq. (4.20) represents the dynamics of a stable system.

This discussion assumes that the M-output current mirror presents no delay. This is not very realistic. Let us assume that the M-output current mirror of Fig. 4.4 has the first-order dynamics defined by the small signal equivalent circuit depicted in Fig. 4.8. Current $I_o(t)$ represents each of the M outputs of this current mirror, and $\sum_j I_{oj}$ its input. The dynamics of such current mirror model in time-domain are given by

$$I_o(t) + \tau_p \dot{I}_o(t) = \sum_j I_{oj}(t) \quad , \quad \tau_p = \frac{C_p}{g_{mp}} \quad (4.23)$$

Eqs. (4.14-4.16) are still valid, but eqs. (4.17-4.19) have a higher order dynamics. Substituting eq. (4.14) and its derivatives into eq. (4.23) results in

$$\tau_p C_c \ddot{v}_{xw}(t) + (C_c + \tau_p G_c)\dot{v}_{xw}(t) + G_c v_{xw}(t) =$$
$$= (G_c v_M - T_w) + \sum_{j=1}^{M} T_j U(v_M - v_{xj}(t)) \qquad (4.24)$$

Subtracting them for two nodes j and w yields,

$$\tau_p C_c [\ddot{v}_{xj}(t) - \ddot{v}_{xw}(t)] + (C_c + \tau_p G_c)[\dot{v}_{xj}(t) - \dot{v}_{xw}(t)] +$$
$$+ G_c [v_{xj}(t) - v_{xw}(t)] = T_w - T_j \qquad (4.25)$$

The solution to this differential equation is

$$v_{xj}(t) - v_{xw}(t) = \frac{T_w - T_j}{G_c} + K_1 e^{-\frac{t}{\tau_p}} + K_2 e^{\frac{t}{\tau_c}} \quad , \quad \tau_c = \frac{C_c}{G_c} \qquad (4.26)$$

where K_1 and K_2 are determined by initial conditions. Consequently, after a few time constants τ_c and τ_p, eq. (4.24) would be given by

$$\tau_p C_c \ddot{v}_{xw}(t) + (C_c + \tau_p G_c)\dot{v}_{xw}(t) + G_c v_{xw}(t) =$$
$$= (G_c v_M - T_w) + \sum_{j=1}^{M} T_j U\left(v_M - v_{xw}(t) - \frac{T_w - T_j}{G_c}\right) \qquad (4.27)$$

If $v_{xw}(t)$ is well above or below $v_M - (T_w - T_j)/G_c$, then the corresponding j-th cell function $T_j U(\cdot)$ will equal either 0 or T_j, respectively. In these cases the term $T_j U(\cdot)$ contributes to the constant term (time independent) of eq. (4.27). On the other hand, if $v_{xw}(t)$ is close to $v_M - (T_w - T_j)/G_c$, term $T_j U(\cdot)$ is close-to-linearly dependent on $v_{xw}(t)$. In this case, its first-order Taylor series expansion is

$$T_j U(\cdot) \approx \frac{1}{2} T_j + T_j U'(0) \left(v_M - \frac{T_w - T_j}{G_c} - v_{xw}(t)\right) \qquad (4.28)$$

This term contributes to the constant term and to the $v_{xw}(t)$ term of eq. (4.27). Summing over all cells obtains

$$\sum_{j=1}^{M} T_j U(\cdot) \approx -S v_{xw}(t) + K, \qquad (4.29)$$

where K and S are constants and $S > 0$ (because T_j, $U'(0) > 0$). Therefore, the poles of eq. (4.27) are the roots of

$$\tau_p C_c s^2 + (C_c + \tau_p G_c)s + (G_c + S) = 0 \qquad (4.30)$$

which always have a negative real part. Consequently, eq. (4.27) converges always to its unique equilibrium point.

4.5 SYSTEM STABILITY FINE ANALYSIS

Performing electrical simulations on the circuit in Section 4.3, verifies that the analysis in Section 4.4 is a good approximation as long as the equilibrium point does not lie in the transition region of any of the M sigmoidal functions $U(\cdot)$. This can only be guaranteed if $\alpha_j = T_j$ and the two largest inputs T_j and T_w are sufficiently different. If $\alpha_j > T_j$ or (with $\alpha_j = T_j$) if two or more inputs T_j are maximum and very similar, the equilibrium point of the system (see Fig. 4.2) will be in the transition region of some sigmoids $U(\cdot)$. In these cases, transistor parasitic elements that have been neglected in the analysis of Section 4.4 may render unstable behavior. Consequently, some kind of compensation is necessary.

Under unstable conditions the system exhibits the following characteristics (observed through electrical simulations with Hspice):

- Only the cells j whose sigmoid functions $U(\cdot)$ must be in their transition region at the equilibrium point are unstable. The rest of the cells behave as if the system had reached its equilibrium point.

- The unstable cells present oscillations (presence of complex conjugate poles).

- In the case of $\alpha_j = T_j$ and with two or more equal maximum inputs, the steady-state oscillating waveforms at these cells become the same, regardless of their initial conditions.

This last observation suggests that a stability analysis could be performed by simply considering one cell in the system, which represents the parallel connection of all unstable cells, as shown in Fig. 4.9(a). On the other hand, since the unstable cells have the equilibrium point in the transition region of their sigmoid $U(\cdot)$, we can linearize these sigmoids for the stability analysis. Therefore, let us consider the small signal equivalent circuit shown in Fig. 4.9(b), where the circuitry comprised by dashen lines represents the parallel of all cells with equal and maximum input. The rest of the circuitry models the M-output current mirror (or set of current mirrors) responsible for distributing the global current I_o among the M cells. The minimum set of dynamic elements needed for the system to present unstable oscillating behavior are parasitic capacitors

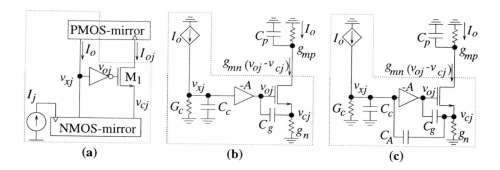

Figure 4.9. (a) Parallel connection of unstable cells, (b) uncompensated small signal equivalent circuit, (c) compensated small signal equivalent circuit.

C_c, C_p, and C_g (observed through electrical simulation). The frequency-domain KCL equations of the linear circuit of Fig. 4.9(b) are

$$
\begin{aligned}
I_o &= V_{xj}(G_c + sC_c) \\
g_{mn}(V_{oj} - V_{cj}) &= g_n V_{cj} - sC_g(V_{oj} - V_{cj}) \\
M_m g_{mn}(V_{oj} - V_{cj}) &= (1 + s\frac{C_p}{g_{mp}})I_o \\
V_{oj} &= -AV_{xj}
\end{aligned}
\quad (4.31)
$$

where M_m is the number of cells with equal and maximum input that are simultaneously active. Routine analysis yields the following third-order polynomial

$$as^3 + bs^2 + cs + d = 0$$

$$
\begin{aligned}
a &= \frac{C_p C_c C_g}{g_{mp}} \\
b &= C_p C_g \frac{G_c}{g_{mp}} + C_p C_c \frac{(g_{mn} + g_n)}{g_{mp}} + C_g C_c \\
c &= C_g G_c + C_p G_c \frac{(g_{mn} + g_n)}{g_{mp}} + C_c(g_{mn} + g_n) \\
d &= (g_{mn} + g_n)G_c + AM_m g_n g_{mn}
\end{aligned}
\quad (4.32)
$$

Since all coefficients a, b, c, and d are positive, the roots of this polynomial have negative real parts if $bc - ad > 0$. Which yields the following stability condition

$$\frac{C_c}{C_p}(g_{mn} + g_n)g_{mp} + \frac{C_c}{C_g}(g_{mn} + g_n)^2 + 2G_c(g_{mn} + g_n) +$$
$$+ \frac{C_p}{C_g}\frac{G_c(g_{mn} + g_n)^2}{g_{mp}} + G_c g_{mp} + \frac{C_p}{C_c}\frac{G_c^2(g_{mn} + g_n)}{g_{mp}} + \frac{C_p}{C_c}G_c^2 >$$
$$> AM_m g_{mn} g_n \tag{4.33}$$

The above expressions have been obtained in an exact manner. That is, no simplifications or approximations have been made. However, in a practical situation some simplifications can be made. As the input node of the current comparator is a high impedance node, conductance G_c (which is the parallel connection of the output conductances of the NMOS and PMOS current mirrors connected to that node) is going to be low when compared to the transconductances g_{mp} and g_{mn}. Thus the approximation $g_{mp}, g_{mn} \gg G_c$ can be made.

With this approximation and considering that parasitic capacitances C_c, C_p and C_g are approximately of the same order of magnitude, the stability condition of eq. (4.33) simplifies to

$$AM_m < \frac{C_c(g_{mn} + g_n)}{g_{mn}g_n}(\frac{g_{mp}}{C_p} + \frac{(g_{mn} + g_n)}{C_g}) \tag{4.34}$$

Note that in the above equation conductance g_n (which is the output conductance of an NMOS current mirror) cannot be neglected against the transconductance g_{mn}. The reason is that each of these current mirrors is receiving an equal and maximum input current T_{MAX} but is delivering to the summing node a lower current T_{MAX}/M_m. This means the output transistor of these current mirrors is in ohmic region and thus it may have a high output conductance g_n.

A more realistic approximation than the one considered in eq. (4.34) would be to consider that capacitance C_g is not of the same order of magnitude than capacitances C_c and C_p. In general, what is going to happen is that capacitance $C_g \ll C_c, C_p$. Considering this approximation together with $g_{mp}, g_{mn} \gg G_c$ the condition of eq. (4.33) results in

$$AM_m < \frac{g_{mn} + g_n}{g_{mn}g_n}\left(\frac{C_c g_{mp}}{C_p} + \frac{g_{mn} + g_n}{C_g}(C_c + \frac{C_p G_c}{g_{mp}})\right) \tag{4.35}$$

The stability conditions expressed in eqs. (4.34) and (4.35) are not easy to satisfy since A must be large for proper operation, M_m may become large, and it is not trivial to make the right hand side of eqs. (4.34) and (4.35) very large.

Stability compensation can be achieved by introducing a capacitor C_A, as shown in Fig. 4.9(c). The frequency domain KCL equations of the linear circuit of Fig. 4.9(c) are

$$\begin{aligned} I_o &= V_{xj}(G_c + sC_c) + sC_A(V_{xj} - V_{cj}) \\ g_{mn}(V_{oj} - V_{cj}) &= g_n V_{cj} - sC_g(V_{oj} - V_{cj}) - sC_A(V_{xj} - V_{cj}) \\ M_m g_{mn}(V_{oj} - V_{cj}) &= (1 + s\frac{C_p}{g_{mp}})I_o \\ V_{oj} &= -AV_{xj} \end{aligned} \quad (4.36)$$

Routine analysis yields the following third-order polynomial

$$as^3 + bs^2 + cs + d = 0$$

$$\begin{aligned} a &= \tau_p[C_c C_g + C_c C_A + (A+1)C_g C_A] \\ b &= \tau_p C_c(g_{mn} + g_n) + \tau_p C_A g_n + \tau_p C_g G_c + C_g C_c + \tau_p G_c C_A + C_c C_A + \\ &\quad + \tau_p(A+1)g_{mn}C_A + (A+1)C_A C_g \\ c &= C_A(g_{mn} + g_n + AM_m g_{mn} + Ag_{mn} + M_m g_{mn}) + \\ &\quad + (\tau_p G_c + C_c)(g_{mn} + g_n) + C_g G_c \\ d &= AM_m g_n g_{mn} + G_c(g_{mn} + g_n) \end{aligned} \quad (4.37)$$

Assuming $A \gg 1$, and $g_{mp}, g_{mn}, g_n \gg G_c$ and $AC_g \gg C_c$ eqs. (4.37) can be simplified to

$$\begin{aligned} a &\approx A\frac{C_g C_A C_p}{g_{mp}} \\ b &\approx AC_A(\frac{g_{mn}}{g_{mp}}C_p + C_g) \\ c &\approx AC_A g_{mn}(1 + M_m) \\ d &\approx AM_m g_n g_{mn} \end{aligned} \quad (4.38)$$

Since all coefficients a, b, c, and d are positive, the roots of this polynomial have negative real parts if $bc - ad > 0$, which yields the following stability condition,

$$(1 + \frac{1}{M_m})C_A > \frac{g_n}{g_{mp}/C_p + g_{mn}/C_g} \quad (4.39)$$

The worst case occurs for very large values of M_m, for which eq. (4.39) reduces to

$$C_A > \frac{g_n}{g_{mp}/C_p + g_{mn}/C_g} \qquad (4.40)$$

Note that now the stability condition does not depend on gain A, and is easier to fulfill. However, now capacitor C_A degrades the settling speed of the system. Capacitor C_A acts as a Miller capacitance. Since the DC-gain from node v_{xj} to node v_{cj} is approximately $-A$ (i.e. the negative of the slope of $U(\cdot)$), there will be an effective Miller capacitance of value $(A+1)C_A$ in parallel with the original C_c capacitor. If the sigmoid is not in its transition region $A \approx 0$, but if the sigmoid is in its transition region A can be very large. Therefore, for compensated cells eq. (4.20) must be changed to

$$[C_c + C_A + U'(v_M - v_{xw})C_A]\dot{v}_{xw} =$$
$$= G_c(v_M - v_{xw}) - T_w + \sum_j T_j U\left(v_M - v_{xw} - \frac{T_w - T_j}{G_c}\right) \qquad (4.41)$$

If the winning cell is in its transition region $U'(v_M - v_{xw}) \neq 0$ and a large capacitance $C_c + (A+1)C_A$ is present at node v_{xw}. Otherwise, $U'(v_M - v_{xw}) = 0$ and the effective capacitance is only $C_c + C_A$.

4.6 EXPERIMENTAL RESULTS

A WTA-MAX system with $M = 10$ competing cells has been designed and fabricated in two different technologies. The first prototype has been integrated in a double-metal single-poly $1.0\mu m$ CMOS technology (ES2), and the other in a double-metal double-poly $2.5\mu m$ CMOS process (MIETEC). Both technologies were available through the European silicon foundry service, EUROCHIP.

If the circuit is going to be used as a MAX circuit, all current mirrors must provide good replication precision. They need to have small systematic errors and small random deviations [Pelgrom, 1989], so that the resulting value of current I_o resembles the maximum among all inputs as much as possible. However, if the circuit is going to be used as a WTA circuit, requirements are not that severe. If inside one single chip, a WTA performs the same even if the current mirrors have appreciable systematic errors. Since systematic errors are common with respect to all inputs, the system can still determine which input is maximum. On the other hand, random mismatch errors in the current mirrors must be kept small because these errors change randomly from one input to another. Reducing random errors implies using larger transistor sizes. Reducing systematic errors implies using more elaborate current mirror topologies that either reduce their output conductance (using cascode [Allen, 1987], regulated cascode [Sackinger, 1990], or gain-boosting [Bult, 1991] techniques), decrease their input impedance [Nairn, 1990b], or both [Serrano-Gotarredona, 1994].

In our case, our ultimate application is WTA operation. Therefore, it is not critical that the final value of I_o be an exact replica of the maximum of the input. Therefore, we used a simple 3-transistor current mirror (without any output conductance or input impedance decreasing technique) for the 2-output NMOS current mirror of each cell. However, we used active input current mirrors [Nairn, 1990b] for the M-output PMOS current mirror and for the extra NMOS assembling current mirror (see Fig. 4.5). These current mirrors assure fixed voltages at their input nodes. This was necessary because if the system is distributed among several chips, the presence of the assembling current mirror would break the symmetry between some of the inputs, making systematic errors affect these inputs differently.

The following presents proper system operation of a WTA circuit in one single chip, in two chips of the same technology, and in two chips each of a different technology. As it will be shown, the DC-behavior of the system is not degraded when the operation is distributed among several chips. In the remainder of this Section we will detail experimental measurements related to the precision of a WTA and its speed response.

Operation Precision

The DC transfer curves of the system have been measured for different input current levels and for different system configurations. Fig. 4.10 shows thirty transfer curves when the competing cells are inside the same chip. Each curve is obtained by randomly selecting a pair of input cells, i and j, applying a constant input current $T_i = T_P$ to the first, and sweeping the input current of the second T_j from $0.9 \times T_P$ to $1.1 \times T_P$. The figure represents the two inverter output voltages, v_{oi} and v_{oj}, versus current T_j. For each pair of cells, i and j, we measure the value of T_j at the point where $v_{oi} = v_{oj}$. Let us call this value T_M. Thirty curves were measured for each value of T_P, resulting in thirty values of T_M. The difference between the mean of these thirty T_M values and T_P is a measure of the systematic error of T_M. Let us call it $\epsilon(T_P)$. The variance of the thirty T_M values represents the random error of T_M. Let us call it $\sigma(T_P)$. In the case of Fig. 4.10, corresponding to a WTA inside one single chip fabricated in the ES2 $1.0\mu m$ CMOS technology with $T_P = 100\mu A$, we measured a random deviation of $\sigma(T_P) = 1.04\%$ and a systematic error of $\epsilon(T_P) = 0.03\%$.

Fig. 4.11 again shows thirty DC transfer curves, where $T_P = 10\mu A$ and the system is built by assembling two chips of the same technology using the set-up illustrated in Fig. 4.5. To obtain these curves, cell i was always chosen among the cells in the first chip, and cell j was always selected from the second chip. Fig. 4.12 shows thirty DC transfer curves for the case $T_P = 500\mu A$, when two chips of different technologies are used. Note that the voltage ranges of v_{oi} and v_{oj} differ for the two chips.

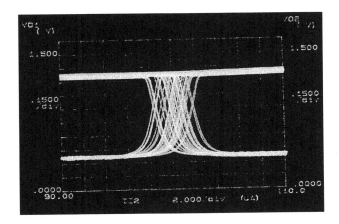

Figure 4.10. Transfer curves of the WTA implemented in a ES2 $1.0\mu m$ chip for an input current level of $100\mu A$ (horizontal scale is $3\mu A/div$ from 90 to $110\mu A$, vertical scale is $0.15V/div$ from $0V$ to $1.65V$).

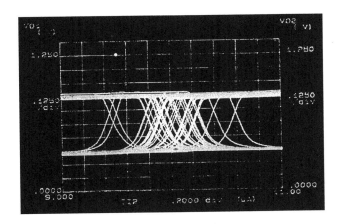

Figure 4.11. Transfer curves when two ES2 $1.0\mu m$ chips are assembled and for an input current level of $10\mu A$ (horizontal scale is $0.2\mu A/div$ from 9 to $11\mu A$, vertical scale is $0.125V/div$ from $0V$ to $1.375V$).

Table 4.2 contains the measured total error (defined as $\sigma(T_P) + \epsilon(T_P)$) for three decades of change in T_P. The table shows results for the cases of WTAs inside one chip, assembled using two chips of the same technology, and assembled with two chips of different technologies. Note that the precision degradation is very small when the system is distributed among two chips, regardless of

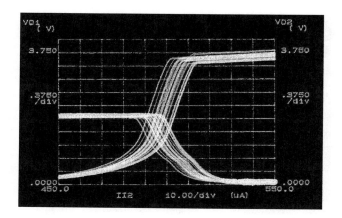

Figure 4.12. Transfer curves when two chips of different technology are assembled and for an input current level of $500\mu A$ (horizontal scale is $10\mu A/div$ from $450\mu A$ to $550\mu A$, vertical scale is $0.375V/div$ from $0V$ to $4.125V$).

whether the chips are of the same technology or not. This is the main advantage of this WTA-MAX circuit with respect to others reported in literature. This is shown in Table 4.3 which depicts the simulation results of another WTA [Choi, 1993]. The input signals for this WTA are voltages that range from 1.5V to 4.5V. As can be seen, there is a significant precision degradation when the WTA is distributed among two chips of different technologies caused by a large increase in the systematic error component [Serrano-Gotarredona, 1995].

Operation Speed

Delay measurements were performed as follows. Only two input signals were made non-zero. Let us call them T_1 and T_2. Current T_1 was made constant and equal to T_{IN}, while current T_2 changed in a pulsed way between values $T_{IN} - 0.5\Delta T_{IN}$ and $T_{IN} + 0.5\Delta T_{IN}$, as shown in Fig. 4.13(a). The pulse starts at time t_{o1} and ends at time t_{o2}. Waveforms v_{o1} and v_{o2} have the shape depicted in Fig. 4.13(b). Four different delay times were measured. For the system response caused by a rising edge in T_2, time t_{d1} is the delay between time t_{o1} and the instant at which voltage v_{o2} crosses the 50% value of its range. Delay t_{d2} is the same for output voltage v_{o1}. For the system response caused by a falling edge in T_2, time t_{d3} is the delay between time t_{o2} and the instant at which voltage v_{o1} crosses the 50% value of its range. Delay t_{d4} is the same for output voltage v_{o2}.

Measurements were performed for T_{IN} values of $10\mu A$, $50\mu A$, $100\mu A$, and $500\mu A$, and for ΔT_{IN} equal to $0.2T_{IN}$ and T_{IN}. Tables 4.4 and 4.5 show the

A HIGH-PRECISION CURRENT-MODE MULTI-CHIP WTA-MAX CIRCUIT

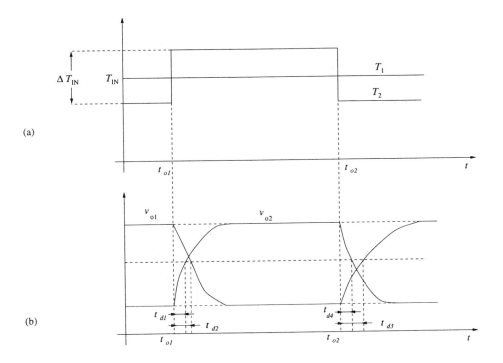

Figure 4.13. (a) Input Signals, (b) Output Waveforms

Table 4.2. Current-Mode WTA Precision Measurements

Technology	number of chips used	T_P			
		$10\mu A$	$100\mu A$	$500\mu A$	$1mA$
ES2_1.0μm	1	2.00%	1.07%	0.58%	0.56%
ES2_1.0μm	2	2.35%	1.03%	0.59%	0.57%
MIETEC_2.4μm	1	1.94%	0.98%	0.70%	0.69%
MIETEC_2.4μm	2	2.15%	1.17%	0.96%	0.87%
MIETEC_2.4μm and ES2_1.0μm	2	2.24%	1.05%	0.73%	0.74%

measured delay times for those cases where the system is inside one single chip. Table 4.6 shows the delay times measured when a WTA is assembled using 2 chips of the ES2 1.0μm process.

Table 4.3. WTA Precision Computations (obtained through Hspice simulations) for the circuit reported in [Choi, 1993]

Technology	number of chips used	v_P 1.5V	2.5V	3.5V	4.5V
ES2_1.0μm	1	0.39%	0.40%	0.41%	1.73%
MIETEC_2.4μm and ES2_1.0μm	2	2.62%	9.65%	13.1%	out of range

Table 4.4. Measured delay times for a one chip WTA in ES2 technology

T_{IN} (μA)	ΔT_{IN} (μA)	ES2_1.0μm t_{d1}	t_{d2}	t_{d3}	t_{d4}
10	2	6.13μs	4.82μs	2.62μs	3.28μs
10	10	1.57μs	1.37μs	1.26μs	1.41μs
50	10	1.29μs	1.02μs	677ns	805ns
50	50	364ns	319ns	308ns	342ns
100	20	943ns	801ns	222ns	254ns
100	100	191ns	167ns	131ns	153ns
500	100	161ns	147ns	125ns	128ns
500	200	59ns	68ns	48ns	9ns

Table 4.5. Measured delay times for a one chip WTA in MIETEC technology

T_{IN} (μA)	ΔT_{IN} (μA)	MIETEC_2.4μm t_{d1}	t_{d2}	t_{d3}	t_{d4}
10	2	3.80μs	2.99μs	3.84μs	5.21μs
10	10	2.08μs	1.82μs	1.98μs	2.28μs
50	10	1.04μs	909ns	1.06μs	1.18μs
50	50	502ns	471ns	455ns	532ns
100	20	594ns	587ns	641ns	637ns
100	100	276ns	249ns	273ns	281ns
500	100	166ns	144ns	112ns	130ns
500	200	104ns	95ns	108ns	102ns

Table 4.6. Measured delay times for a two-chips WTA

T_{IN}	ΔT_{IN}	\multicolumn{4}{c}{$ES2\_1.0\mu m$}
		t_{d1}	t_{d2}	t_{d3}	t_{d4}
$10\mu A$	$2\mu A$	$16.8\mu s$	$6.50\mu s$	$5.40\mu s$	3.40μ
$10\mu A$	$10\mu A$	$3.2\mu s$	$2.35\mu s$	$1.86\mu s$	$1.80\mu s$
$100\mu A$	$20\mu A$	$470ns$	$480ns$	$750ns$	$590ns$
$100\mu A$	$100\mu A$	$235ns$	$230ns$	$270ns$	$240ns$
$500\mu A$	$100\mu A$	$154ns$	$134ns$	$375ns$	$150ns$
$500\mu A$	$200\mu A$	$150ns$	$104ns$	$136ns$	$110ns$

5 AN ART1/ARTMAP/FUZZY-ART/FUZZY-ARTMAP CHIP

In this Chapter we present a new design approach for a *Second Generation* ART chip. This chip would be able to emulate the ART1, Fuzzy-ART, ARTMAP, and Fuzzy-ARTMAP architectures, and using both choice functions possibilities: subtraction and division for computing the T_j terms. The synaptic cells for this chip operate in weak inversion, so that chip size can be scaled up easily without worrying too much about power dissipation limitations. Also, the synaptic cells present a very high area density: we estimate that for a $1cm^2$ chip in a $0.5\mu m$ CMOS process with 3 metal layers, an ART1 (or simplified ARTMAP) system with $N = 385$ input nodes and $M = 430$ category nodes can be implemented; or if used as Fuzzy-ART (or simplified Fuzzy-ARTMAP) it could be of $N = 77$ input nodes and $M = 430$ category nodes.

This cell density is achieved thanks to the use of a circuit element called here the "*Current Source Flip-Flop*". It is a very simple circuit inspired on other works [Pouliquen, 1997], [Lyon, 1987] which implements a flip-flop that outputs a specified current or a null current, depending on its stored state. This cell can be used for an ART1 synaptic cell, or n of them can be grouped to form an n-bit Fuzzy-ART cell. For the chip we are describing in this Chapter $n = 5$ has been chosen which renders a precision of $1/32$ or approximately 3%.

The description of the chip is divided into two main parts. First, the description of the basic processing synaptic cell and second, the description of the peripheral cells. The synaptic cell should be of minimum area and power consumption, so that a very high density system can result with reasonable power consumption. Note that the final area and power consumption of the system depends quadratically on those of the synaptic cell. The peripheral cells affect only linearly to the final system area and power consumption, and consequently their designs can be more relaxed in this sense. In conclusion, it is a good practice to simplify the synaptic cell at the expense of complicating the peripheral cells. One drawback of this is that many times it turns out that peripheral cells result to be very thin but extremely long, which makes them difficult to be laid out.

5.1 THE SYNAPTIC CELL

The cell described in this Section is a Fuzzy-ART cell capable of realizing the following operations:

1. Compute the analog value of a 5-bit stored pattern

$$z_{ij} = \sum_{n=1}^{5} z_{ij}^{(n)} \frac{1}{2^n} \tag{5.1}$$

This value is generated as a current by the cell so that the l_1 norm of vector $\mathbf{z}_j$

$$|\mathbf{z}_j| = |(z_{1j}, z_{2j}, \ldots, z_{Nj})| = \sum_{i=1}^{N} z_{ij} \tag{5.2}$$

can be computed by simply adding these currents for all cells in the same row.

2. In order to compute the fuzzy minimum between an input vector $\mathbf{I}$ and the template vector $\mathbf{z}_j$ of a stored category $|\mathbf{I} \wedge \mathbf{z}_j|$, each cell has to find this minimum for one of the components

$$I_i \wedge z_{ij} \tag{5.3}$$

Both I_i and z_{ij} are stored on-chip as 5-bit digital words.

3. Be able to update the stored template

$$\mathbf{z}_J(new) = \mathbf{I} \wedge \mathbf{z}_J(old) \tag{5.4}$$

which means that the cells to be updated must perform

$$z_{iJ}(new) = I_i \wedge z_{iJ}(old) \tag{5.5}$$

As will be seen later this update operation will be carried out by the *Top* peripheral cells.

This Fuzzy-ART cell, when biased and used by the peripheral cells in a different manner, performs the function of five ART1 cells. In this case, the operations each ART1 cell performs are:

1. Generate a current controlled by the stored bit z_{ij}. If $z_{ij} = 0$ the current should be zero, and if $z_{ij} = 1$ the current is a given constant.

2. Generate a current controlled by the AND operation of stored bit z_{ij} and input pattern pixel I_i,

$$I_i z_{ij} \tag{5.6}$$

3. Update the stored z_{ij} value as

$$z_{ij}(new) = I_i z_{ij}(old) \tag{5.7}$$

This is achieved by the use of a very simple circuit element which we like to call the "*Current Source Flip-Flop*".

The Current Source Flip-Flop

Fig. 5.1 shows the basic concept behind the *Current Source Flip-Flop* circuit. Two current sources of equal value I_u are connected with two NMOS transistors $M1$ and $M2$ in a flip-flop configuration. The sources of the two NMOS transistors are connected to two lines tied to a fixed voltage V_S by peripheral voltage sources. Depending on the state of the flip-flop the currents injected into these lines change: if $z_{ij} = 1$ ($\overline{z_{ij}} = 0$), $M1$ is ON, $M2$ is OFF, and the circuit injects a current I_u into line z_j^+, while no current is injected into line z_j^-; conversely, if $z_{ij} = 0$ ($\overline{z_{ij}} = 1$), $M1$ is OFF, $M2$ is ON, and the circuit injects current I_u into line z_j^- but no current into line z_j^+.

If N of these *Current Source Flip-Flops* are connected horizontally sharing lines z_j^+ and z_j^-, the total current injected into these lines would be

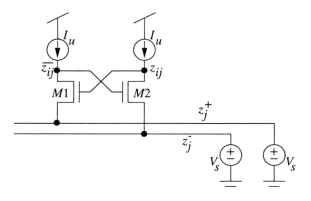

Figure 5.1. Conceptual Diagram of Current Source Flip-Flop

$$I(z_j^+) = \sum_{i=1}^{N} z_{ij} I_u = I_u |\mathbf{z}_j| \tag{5.8}$$

$$I(z_j^-) = \sum_{i=1}^{N} \overline{z_{ij}} I_u = I_u (N - |\mathbf{z}_j|)$$

All *Current Source Flip-Flops* in the same row can be set to '1' by momentarily lowering voltage V_S for line z_j^+:

1. If for a given cell $z_{ij} = 0$ and $\overline{z_{ij}} = 1$ ($M1$ OFF and $M2$ ON), lowering voltage V_S for line z_j^+ will turn $M1$ ON and the voltage at node $\overline{z_{ij}}$ will come down. If the voltage for node z_j^+ is sufficiently low, voltage at node $\overline{z_{ij}}$ will go sufficiently low to turn OFF transistor $M2$ and thus changing the state of the flip-flop. If voltage at node z_j^+ returns now to its original value, it will be $z_{ij} = 1$ and $\overline{z_{ij}} = 0$ ($M1$ ON and $M2$ OFF).

2. If for a given cell $z_{ij} = 1$ and $\overline{z_{ij}} = 0$ ($M1$ ON and $M2$ OFF), lowering voltage V_S for line z_j^+ will not alter the state of the flip-flop.

On the other hand, if voltage for line z_j^- is momentarily changed to a sufficiently low voltage value the state for all *Current Source Flip-Flops* in this row will be set to $z_{ij} = 0$ and $\overline{z_{ij}} = 1$.

A Fuzzy-ART Current Source Memory Cell

Grouping n *Current Source Flip-Flops* with binarily weighted current sources yields an n-bit *Current Source Memory Cell*. This is shown in Fig. 5.2. In this

AN ART1/ARTMAP/FUZZY-ART/FUZZY-ARTMAP CHIP

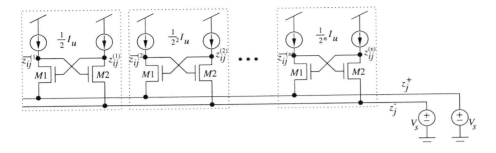

Figure 5.2. Grouping of Current-Source Flip-Flops to assemble an n-bit Current-Source Memory Cell suitable for Fuzzy-ART Synaptic Cells

case, the current injected into lines z_j^+ and z_j^- would be

$$I(z_j^+) = I_u \sum_{l=1}^{n} z_{ij}^{(l)} \frac{1}{2^l} = I_u z_{ij} \quad (5.9)$$

$$I(z_j^-) = I_u \sum_{l=1}^{n} \overline{z_{ij}}^{(l)} \frac{1}{2^l} = I_u \overline{z_{ij}}$$

where z_{ij} and $\overline{z_{ij}}$ may take values between 0 and $1 - 1/2^n$ in steps of $1/2^n$ and satisfy

$$\overline{z_{ij}} + z_{ij} = \sum_{l=1}^{n} \frac{1}{2^l} = 1 - \frac{1}{2^n} \quad (5.10)$$

Connecting N Fuzzy-ART *Current-Source Memory Cells* in a row sharing lines z_j^+ and z_j^- enables the generation of currents

$$I(z_j^+) = I_u \sum_{i=1}^{N} z_{ij} = I_u |\mathbf{z}_j| \quad (5.11)$$

$$I(z_j^-) = I_u \sum_{i=1}^{N} \overline{z_{ij}} = I_u \left[N(1 - \frac{1}{2^n}) - |\mathbf{z}_j| \right]$$

where $\mathbf{z}_j$ is an N component vector with each component represented by an n-bit digital word (from now on it is assumed that $n = 5$).

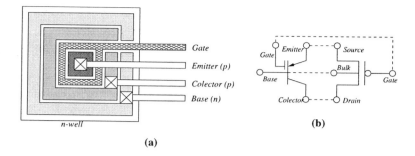

Figure 5.3. (a) Physical Layout for CMOS Compatible Lateral Bipolar (pnp) Transistor, (b) Equivalent Schematic Representation of the resulting Parallel Connection of Bipolar pnp and PMOS.

Implementation of the Current Sources

Current sources I_u of Fig. 5.1 or $\frac{I_u}{2^i}$ of Fig. 5.2 should be generated in such a way that mismatch in reference value I_u is as small as possible while not penalizing too much area consumption. One possibility is to use CMOS compatible lateral bipolar transistors [Arreguit, 1989] whose physical layout for an n-well process is shown in Fig. 5.3(a). It is known that the mismatch behavior for these transistors is much better than for equivalent MOS transistors [Pan, 1989]. The schematic representation of this structure is shown in Fig. 5.3(b) where a bipolar pnp and a PMOS transistor appear connected in parallel[1]. To activate the bipolar transistor the gate voltage at terminal G should be biased above the positive power supply (usually $1V$ above) of the chip, so that the PMOS transistor is shut off and free holes below the gate are completely eliminated to enhance bipolar operation mode. Using this idea the *Current Source Flip-Flop* of Fig. 5.1 would result in the circuit shown in Fig. 5.4(a). Note that transistors $Q1$ and $Q2$ share their Emitter and Base terminals, which allows for the compact layout depicted in Fig. 5.4(b).

Changing voltage V_{BASE} controls the value of unit current I_u. If all cells in the same row have the same I_u (as it would be the case for an ART1 chip) then voltage V_{BASE} is the same for all *Current Source Flip-Flops* and the n-well can be shared by all cells. This helps tremendously to produce a very dense layout.

However, for the case of the Fuzzy-ART *Current Source Memory Cell* the unit currents for each flip-flop are different. Furthermore, the ratio between maximum and minimum currents may become significantly large depending on the number of bits one wants to implement. In this case, it is better to control for each flip-flop the current through the Emitter voltage of transistors Q_1 and

---

[1]Strictly speaking, the layout of Fig. 5.3(a) has 3 bipolar and one PMOS transistors [Arreguit, 1989].

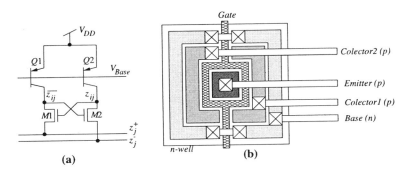

Figure 5.4. (a) Current Source Flip-Flop with Bipolar pnp Transistors as Current Sources, (b) Physical Layout of a Two-Collector Lateral Bipolar pnp Structure.

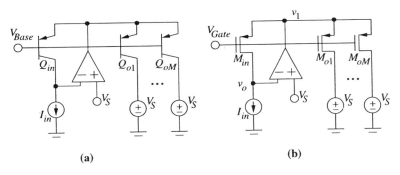

Figure 5.5. Source/Emitter driven p-type current mirrors, (a) bipolar and (b) MOS versions

Q_2 while maintaining the same Base voltage for all flip-flops. These emitter voltages can be controlled from the periphery of the chip and can be shared by all flip-flops in the same column. For this purpose the current mirror described next results very convenient.

A Source/Emitter Driven Active Input Current Mirror

Fig. 5.5 shows a bipolar and a MOS version for a p-type Emitter/Source driven active input current mirror [Serrano-Gotarredona, 1998b]. These structures assure that Collector (or Drain) voltages of input and output transistors are the same: the output transistors are the current source transistors Q_1 or Q_2 of Fig. 5.4(a) whose Collectors will be connected to voltage V_S. Consequently, the use of this current mirror to generate the Emitter voltage of all synaptic current sources eliminates the Early voltage effect, which is very high for minimum size lateral bipolars [Arreguit, 1989].

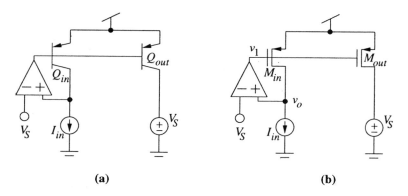

Figure 5.6. Conventional Active-Input Current Mirror, (a) bipolar and (b) MOS versions

The advantages of the Emitter/Source driven active input current mirrors of Fig. 5.5 over the conventional active input current mirror, shown in Fig. 5.6 are:

1. As mentioned earlier, it allows to share the Base voltage (in the bipolar case) for all current mirrors on the same chip, independently of the current value to be mirrored. This fact allows all current mirrors to share their Base diffusion (or well), which yields a much higher density layout.

2. From a circuit stability point of view, the Emitter/Source driven current mirror is much more benevolent than the conventional active-input current mirror. Fig. 5.7 shows the small signal equivalent circuits of the current mirror input stage (voltage amplifier and transistor M_{in}) for the MOS versions in Fig. 5.5(b) and Fig. 5.6(b), assuming the differential input voltage amplifier can be modeled by a single pole behavior (g_{m1}, g_{o1} and C_{p1}). Capacitance C_p models the parasitic capacitor at the input node of the mirrors and g_{m2}, g_{o2}, C_{gd2} are small signal parameters for transistor M_{in}. For the conventional structure (Fig. 5.7(b)) the resulting characteristics equation is

$$s^2[C_{p1}C_p + C_{gd2}(C_{p1} + C_p)] + s[g_{o1}(C_p + C_{gd2}) + g_{o2}(C_{p1} + C_{gd2}) + \\ + C_{gd2}(g_{m2} - g_{m1})] + [g_{o1}g_{o2} + g_{m1}g_{m2}] = 0 \quad (5.12)$$

By reducing I_u, g_{m2} and g_{o2} are made smaller, and may make the s coefficient negative, introducing instability and the need for compensation. On the contrary, for the circuit in Fig. 5.7(a) the characteristics equation is

$$s^2 C_p C_{p1} + s[(g_{o1} + g_{o2} + g_{m2})C_p + g_{o2}C_{p1}] + \\ + [g_{o1}g_{o2} + g_{m1}(g_{o2} + g_{m2})] = 0 \quad (5.13)$$

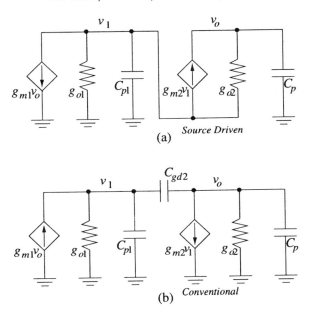

Figure 5.7. Small Signal Equivalent Circuits for (a) Source Driven Active Input, and (b) Conventional Active Input Current Mirrors

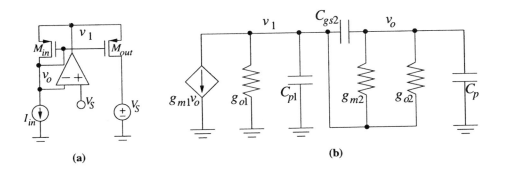

Figure 5.8. Alternative Source Driven Active Input Current Mirror

which is such that its s coefficient never becomes negative. This structure, however, may present poor phase margins and become unstable if the voltage amplifier has high gains and more than one poles. For these cases, the alternative structure of Fig. 5.8(a) can be used. Assuming a single pole behavior for the voltage amplifier, the small signal equivalent circuit shown in Fig. 5.8(b) results for the input stage (transistor M_{in} and amplifier). The characteristics equation for this circuit is

$$s^2[C_{p1}(C_p + C_{gs2}) + C_p C_{gs2}] + s[C_{gs2}(g_{m1} + g_{o1}) + C_{p1}(g_{m2} + g_{o2}) +$$
$$+ C_p(g_{o1} + g_{o2} + g_{m1} + g_{m2})] + (g_{m2} + g_{o2})(g_{m1} + g_{o1}) = 0$$
(5.14)

Note that since M_{in} is connected in a passive diode configuration around the negative feedback loop of the voltage amplifier, the circuit will remain stable independently of the type of amplifier used (assuming, of course, the amplifier itself is compensated).

3. The same physical circuit (Fig. 5.5) can be made to operate either in bipolar mode or in MOS mode by properly controlling the Gate and Bulk (or Base) voltages: for bipolar mode Gate voltage must be set above V_{DD} and well (or Base) voltage to an intermediate value (at least 0.8V below the positive saturation voltage of the amplifier), while for MOS operation well voltage should be at V_{DD} and Gate voltage intermediate (at least one PMOS threshold voltage below the positive saturation voltage of the amplifier for strong inversion, or less for weak inversion operation).

The Complete Fuzzy-ART Synaptic Cell

The current source multi-bit memory cell principle of Fig. 5.2 only allows to store a 5-bit weight z_{ij} and to generate an analog current proportional to this weight. Besides this, for proper Fuzzy-ART operation we need to be able to generate as well the fuzzy-min term

$$I_i \wedge z_{ij} \tag{5.15}$$

Fig. 5.9(a) shows the complete cell. Each *Current Source Flip-Flop* has an extra branch to inject current $\overline{z_{ij}}^{(k)} \frac{I_u}{2^k}$ into node '*local*' (assuming lines $\overline{I_i}^{(k)}$ are set high). Therefore, the total current injected into this node is proportional to $\overline{z_{ij}} = (1 - 1/2^n) - z_{ij}$. Since now 3 bipolar transistors (Q_0, Q_1, Q_2) are needed, their layout can be as shown in Fig. 5.9(b). Transistors Q_0 and Q_2 should match well and therefore they should use the two upper Collectors in Fig. 5.9(b). Transistor Q_1 is only needed to assure storage and thus its current is not of critical importance: it can use the lower Collector. Transistor Q_I is the output transistor of an Emitter driven current mirror whose input stage is on the periphery and which drives a current proportional to

$$\overline{I_i} = (1 - 1/2^n) - I_i \tag{5.16}$$

Voltage V_{FA} is set high for Fuzzy-ART mode, and low for ART1 mode. When voltage V_{FA} is connected to positive power supply, transistor M_7 is OFF and

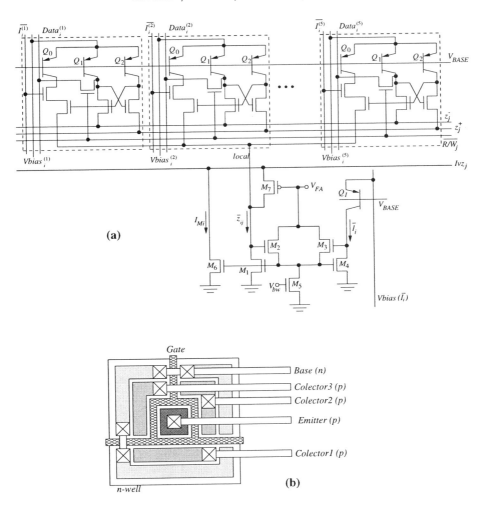

Figure 5.9. Complete Circuit Schematic of ART1/Fuzzy-ART Synaptic Cell

transistors $M_1 - M_5$ constitute a 2-input WTA circuit [Lazzaro, 1989]. Therefore, transistor M_6 will mirror the maximum current among those flowing through M_1 ($\overline{z_{ij}}$) or M_4 ($\overline{I_i}$),

$$I_{Mi} = max\{\overline{I_i}, \overline{z_{ij}}\} = \overline{I_i} \vee \overline{z_{ij}} \qquad (5.17)$$

Consequently, node Ivz_j, which is shared by all Fuzzy-ART cells in the same row sinks a current proportional to

$$I(Ivz_j) = \sum_{i=1}^{N} \overline{I_i} \vee \overline{z_{ij}} = \sum_{i=1}^{N} \left[\left(1 - \frac{1}{2^n}\right) - I_i\right] \vee \left[\left(1 - \frac{1}{2^n}\right) - z_{ij}\right] =$$

$$= \sum_{i=1}^{N} \left[\left(1 - \frac{1}{2^n}\right) - I_i \wedge z_{ij}\right] = N\left(1 - \frac{1}{2^n}\right) - |\mathbf{I} \wedge \mathbf{z}_j| \qquad (5.18)$$

Since term $N(1 - 1/2^n)$ is constant and equal for all rows, the current flowing through node Ivz_j can be used to obtain $|\mathbf{I} \wedge \mathbf{z}_j|$ which is needed to compute the choice function terms T_j for the Fuzzy-ART algorithm.

Line $\overline{R/W_j}$ is used to connect nodes $z_{ij}^{(k)}$ to lines $Data_i^{(k)}$ so that the value of $z_{ij}^{(k)}$ can be either read or set. For reading $z_{ij}^{(k)}$ care must be taken to not alter its stored value. Therefore, during a read cycle, unit currents I_u biasing the flip-flops should be momentarily increased.

The same cell of Fig. 5.9 can be used as n ART1 cells by connecting voltage V_{FA} to ground and bias voltage V_{bw} to power supply. This will connect line '*local*' to line Ivz_j directly and turn OFF transistors M_1, M_4 and M_6. Also, in ART1 mode, lines $\overline{I_i}^{(k)}$ are not connected any more to the power supply voltage, but to the corresponding complemented value of an input pattern binary pixel. Under these circumstances, the current injected into line Ivz_j is now proportional to

$$I(Ivz_j) = \sum_{i=1}^{nN} \overline{z_{ij} I_i} = \sum_{i=1}^{nN}(1 - z_{ij})(1 - I_i) = nN - |\mathbf{z}_j| - |\mathbf{I}| + |\mathbf{I} \cap \mathbf{z}_j| \quad (5.19)$$

which can be used to extract $|\mathbf{I} \cap \mathbf{z}_j|$ if $|\mathbf{z}_j|$ and $|\mathbf{I}|$ are available. This task will be performed by peripheral circuits.

The cell of Fig. 5.9 has been laid out for a 3-metal $0.5\mu m$ CMOS process and occupies an area of $19.95\mu m \times 101.15\mu m$. This layout is shown in Fig. 5.10.

5.2 PERIPHERAL CELLS

The function of the peripheral cells is to properly bias the synaptic cells and to extract from them the appropriate information, as well as to read and update the stored memory values.

Fig. 5.11 shows an approximate floorplan for a $1cm^2$ chip. The chip consists of 77×430 synaptic cells of size $19.95\mu m \times 101.15\mu m$, 77 bottom cells each of size $101.15\mu m \times 825.30\mu m$, 77 top cells of size $101.15\mu m \times 248.15\mu m$, 430 right cells each of size $19.95\mu m \times 927.5\mu m$, a 430-input Winner-Takes-All circuit of

AN ART1/ARTMAP/FUZZY-ART/FUZZY-ARTMAP CHIP 129

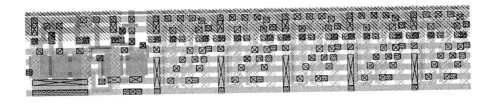

Figure 5.10. Layout of ART1/Fuzzy-ART Synaptic Cell

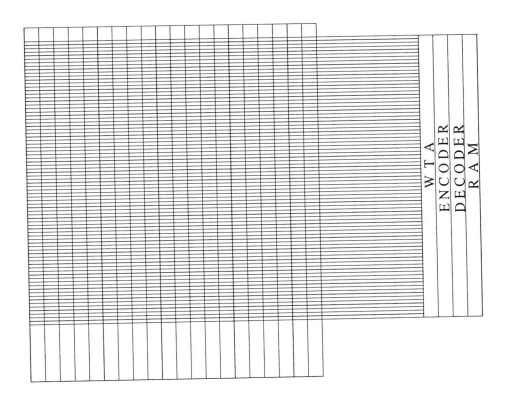

Figure 5.11. Floorplan of ART1/Fuzzy-ART/ARTMAP/Fuzzy-ARTMAP Chip

size $315.70\mu m \times 8578.5\mu m$, an Encoder/Decoder pair of size $151\mu m \times 8578.5\mu m$ and a RAM circuit of size $357.7\mu m \times 8578.5\mu m$.

The functions of the peripheral cells are the following:

1. **Bottom Cells:**

   (a) Read input pattern **I** and latch it into a register, while the next pattern can be read in.

   (b) Provide the appropriate Emitter voltages for all *Current-Source Flip-Flops* on the same column, for both modes of operation: ART1 and Fuzzy-ART.

   (c) Generate the currents $|\mathbf{I}|I_u$ and nNI_u (ART1) or $N(1-1/2^n)I_u$ (Fuzzy-ART) needed by other peripheral cells.

2. **Top Cells:**

   (a) Read out the stored category vector $\mathbf{z}_j$ either for the winning category or for a preselected category.

   (b) Learning Rule Implementation: do the component-wise minimum between vectors **I** and $\mathbf{z}_j$ digitally and store this value in the preselected category vector.

   (c) Alternative Learning Rule: if one wants to perform another operation between vectors **I** and $\mathbf{z}_j$, vector $\mathbf{z}_j$ can be read out of the chip, perform whatever operation is desired externally and feed the result in again to update the corresponding category vector $\mathbf{z}_j$.

3. **Right Cells:**

   (a) Provide low impedance current sourcing/sinking nodes for lines z_j^+, z_j^-, and Ivz_j.

   (b) Generate the appropriate T_j terms using the currents supplied by the synaptic cells and bottom cells. These currents are either in the form

$$T_j = \frac{|\mathbf{I} \wedge \mathbf{z}_j|}{\alpha + |\mathbf{z}_j|} \quad \text{(Fuzzy-ART)} \tag{5.20}$$

$$T_j = \frac{|\mathbf{I} \cap \mathbf{z}_j|}{\alpha + |\mathbf{z}_j|} \quad \text{(ART1)}$$

for division based choice functions, or

$$T_j = L_A|\mathbf{I} \wedge \mathbf{z}_j| - L_B|\mathbf{z}_j| + L_M \quad \text{(Fuzzy-ART)} \tag{5.21}$$
$$T_j = L_A|\mathbf{I} \cap \mathbf{z}_j| - L_B|\mathbf{z}_j| + L_M \quad \text{(ART1)}$$

for difference based choice functions.

(c) Check if the vigilance criterion is satisfied

$$\rho|\mathbf{I}| \leq |\mathbf{I} \wedge \mathbf{z}_j| \quad \text{(Fuzzy-ART)} \tag{5.22}$$
$$\rho|\mathbf{I}| \leq |\mathbf{I} \cap \mathbf{z}_j| \quad \text{(ART1)}$$

and inhibit or not its corresponding choice function term T_j.

(d) Provide access to intermediate processing stages for test purposes.

4. **Winner-Takes-All:** The function of the Winner-Takes-All (WTA) is to select the maximum T_j among all categories. Since the number of categories can be up to 430, it has been implemented as a cascade of two WTA circuits to improve its discriminatory ability.

5. **Encoder:** Since there is a high number of category nodes, the encoder transforms the winning category number J into a binary word which can be sent out of the chip.

6. **Decoder:** For test purposes it is convenient to be able to access a predetermined category (not necessarily the winning one). An external digital word selects through the decoder the category line that one wants to activate.

7. **RAM:** This RAM provides the chip with the functionality to emulate either ARTMAP or Fuzzy-ARTMAP behavior.

In what follows we briefly describe these peripheral cells.

Bottom Cells

The bottom cells contain a 5-bit shift register which connects serially to all other bottom cells. This way input pattern **I** is loaded serially into the chip. Once the pattern is loaded its value is copied to the 5-bit register R. The complement of the stored value is sent to the top cells in the same column through lines $\overline{I1'}, \ldots \overline{I5'}$. The rest of the circuit performs differently depending on the operation mode.

Fuzzy-ART Mode. In Fuzzy-ART mode lines $\overline{I1}, \ldots \overline{I5}$ are always high. The bottom cell also contains current sources[2] BCS and BCS'. In Fuzzy-ART operation mode, circuit BCS provides 5 binarily weighted current sources $I^{(1)}, \ldots I^{(5)}$ such that

$$I^{(k)} = I_u \frac{1}{2^k} \tag{5.23}$$

---

[2] BCS stands for *"Bottom Current Source"*

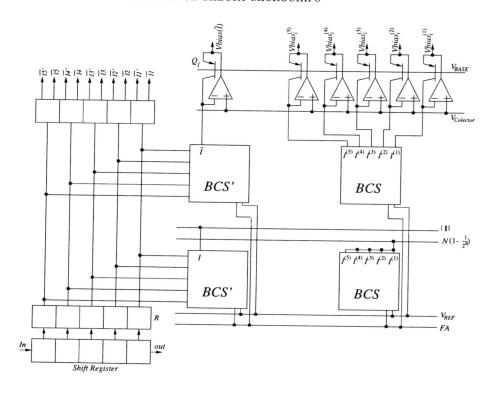

Figure 5.12. Schematic Circuit for Bottom Cells

Unit current I_u is controlled through voltage V_{REF}. Circuit BCS' contains a BCS circuit with some additional switches which allow to generate any combination of currents $I^{(1)}$ through $I^{(5)}$, controlled by the 5-bit digital word latched in register R. In the case of the bottom BCS' circuit of Fig. 5.12 the output current is proportional to I_i (the 5-bit value of the $i-th$ pixel of input pattern **I**),

$$I_i = I_u \sum_{k=1}^{5} \frac{1}{2^k} I_i^{(k)} \qquad (5.24)$$

where I_u is the unit current (common for the whole chip) and $I_i^{(k)}$ are the bits that compose the 5-bit digital word representing I_i. This current contributes to the computation of term $|\mathbf{I}|$ needed by the *Right* peripheral cells. In the case of the top BCS' circuit of Fig. 5.12 the output current is the complementary,

$$\overline{I_i} = \left[\left(1 - \frac{1}{2^n}\right) - I_i\right] I_u \qquad (5.25)$$

This current is fed to the input of an Emitter-Driven active input current mirror and the resulting emitter voltage is shared by all emitter nodes $V_{bias}(\overline{I_i})$ (see Fig. 5.9) on the same column.

Regarding the BCS current sources in Fig. 5.12, the bottom one is used to generate

$$N\left(1 - \frac{1}{2^n}\right) I_u = NI_u \sum_{k=1}^{n} I^{(k)} \qquad (5.26)$$

needed by the *Right* peripheral cells, and the top one generates the binarily weighted currents for the *Current-Source Flip-Flops* of the synaptic cells.

Besides all this, the *Bottom* cells also include some extra circuitry that short circuits Emitter and Base terminals of transistor Q_I in Fig. 5.12 for the case the 5-bit pixel $\overline{I_i}$ is identically zero.

ART1 Mode. When the chip operates in ART1 mode, circuits BCS and BCS' generate a set of equal currents instead of binarily weighted. Bottom circuit BCS' contributes to generate

$$I_u |\mathbf{I}| = I_u \sum_{i=1}^{nN} I_i \qquad (5.27)$$

while bottom circuit BCS contributes to

$$\sum_{i=1}^{nN} I_u = nNI_u \qquad (5.28)$$

Top circuit BCS' is not used by the synaptic cells in ART1 mode, and top circuit BCS generates 5 equal currents used to bias the emitters of the *Current Source Flip-Flops* of the complete 5-segment column.

Top Cells

The main function of these cells is to upload the 5-bit z_{ij} value for column i and for a preselected category (either the winning one, or one selected externally),

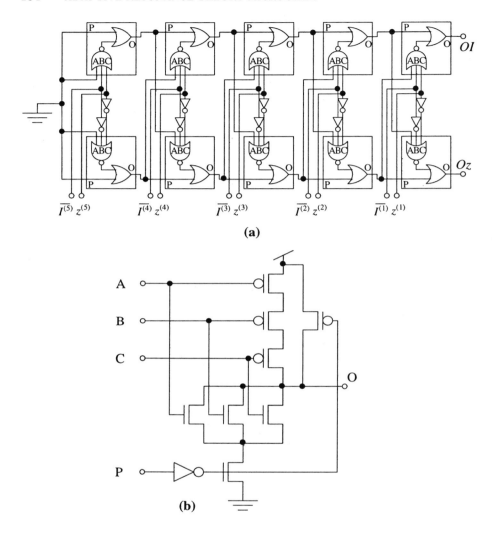

Figure 5.13. Simplified Circuit Diagram of Top Cells

upload the 5-bit I_i value for this column, determine digitally which is minimum and write the result back to the corresponding synaptic cell. Once z_{ij} and I_i are read and latched, a circuit like the one shown in Fig. 5.13(a) determines the minimum. Output OI is high if I_i is larger than z_{ij}, while Oz is high in the other case. If $I_i = z_{ij}$ then both outputs are low.

The processing is based on 10 processing cells whose circuit diagram is shown in Fig. 5.13(b). Each cell compares the $k-th$ bit of I_i and z_{ij} ($I^{(k)}$ and $z^{(k)}$). If they are equal, output O will be low and the next bit needs to be compared. If they are not equal, output O goes high for either the bottom row in Fig. 5.13(a) (if $I^{(k)} = 0$ and $z^{(k)} = 1$), or the top row (if $I^{(k)} = 1$ and $z^{(k)} = 0$). Once one of the O outputs is high, the consequent outputs for the same row will remain high and those of the other row will be kept low.

Additional circuitry, not shown in Fig. 5.13, is included in the top cells to read out the z_{ij} values, write in the new ones and propagate them to the synaptic cells. Also, some extra signal conditioning circuits are needed to adapt the high and low levels in the synaptic flip-flops to the full range needed for digital processing.

Right Cells

Lines z_j^- (see Fig. 5.9) are tied directly to current sinking voltage sources of value V_S, and the currents flowing through them are not processed. On the other hand, lines z_j^+ and Ivz_j are connected to active input current mirrors that clamp their voltages to V_S, and their currents are properly processed. Depending on whether the chip operates in ART1 or Fuzzy-ART mode, the currents flowing through these lines are different and require different processing. In Fuzzy-ART mode the currents are

$$I(z_j^+) = +|\mathbf{z}_j|I_u \qquad (5.29)$$
$$I(Ivz_j) = -\left\{N(1 - \frac{1}{2^n}) - |\mathbf{z}_j \wedge \mathbf{I}|\right\} I_u$$

while in ART1 mode they are

$$I(z_j^+) = +|\mathbf{z}_j|I_u \qquad (5.30)$$
$$I(Ivz_j) = +\left\{nN - |\mathbf{z}_j| - |\mathbf{I}| + |\mathbf{z}_j \cap \mathbf{I}|\right\} I_u$$

Consequently, the circuits that process these currents need to be reconfigured depending on whether the chip operates in ART1 or Fuzzy-ART mode, or whether the choice functions are based on division or subtraction.

Fig. 5.14(a) shows the configuration needed for Fuzzy-ART with subtraction based choice functions, and Fig. 5.14(b) for division based ones. For subtraction based choice functions, two current mirrors (one of gain $\alpha > 1$ and the other of unity gain) are required as well as three current sources: $N(1 - 1/2^n)I_u$ (this current is generated by the *Bottom* cells), I_{M1}, and I_{M2}, whose values are not critical as long as they match well for all *Right* cells. The final current flowing into the corresponding vigilance circuit is

$$T_j = I_{M2} - I_{M1} + \alpha|\mathbf{I} \wedge \mathbf{z}_j|I_u - |\mathbf{z}_j|I_u \qquad (5.31)$$

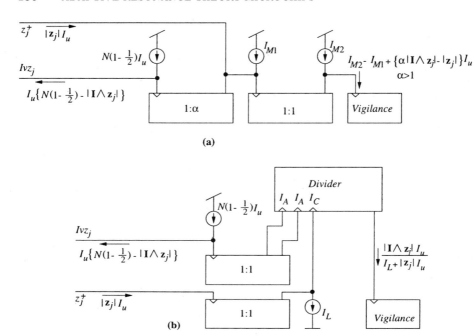

Figure 5.14. Circuit Configuration of Right Cells for Fuzzy-ART Operation Mode with (a) subtraction based and (b) division based choice functions

For division based choice functions two unity gain current mirrors, two current sources of value $N(1-1/2^n)I_u$ and I_L, and a division circuit are required. This results in a current flowing into the vigilance circuit of

$$T_j = \frac{|\mathbf{I} \wedge \mathbf{z}_j|I_u}{I_L + |\mathbf{z}_j|I_u} \qquad (5.32)$$

The current division circuit is explained later.

In ART1 mode the processing circuits change. For difference based choice functions the circuit, shown in Fig. 5.15(a), requires two unity gain current mirrors (one N and one P), an N mirror of gain $\alpha' < 1$, and three current sources of values NnI_u, $|\mathbf{I}|I_u$ and I_{M3}. This provides the current

$$T_j = |\mathbf{I} \cap \mathbf{z}_j|I_u + (\alpha'-1)|\mathbf{z}_j|I_u + I_{M3} \ , \ \alpha' < 1 \qquad (5.33)$$

In the case of division based ART1 processing the circuit of Fig. 5.15(b) can be used. This requires two unity gain current mirrors, five current sources: two of

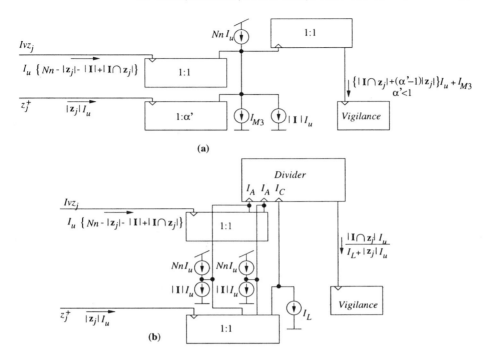

Figure 5.15. Circuit Configuration of Right Cells for ART1 Operation Mode with (a) subtraction based and (b) division based choice functions

value nNI_u, two of value $|\mathbf{I}|I_u$, and one of value I_L, and a division circuit. The output current in this case is

$$T_j = \frac{|\mathbf{I} \cap \mathbf{z}_j|I_u}{I_L + |\mathbf{z}_j|I_u} \tag{5.34}$$

Division Circuit

The division circuit, whose complete schematic is shown in Fig. 5.16(a), performs a current division operation either if transistors operate in strong inversion or in weak inversion. The circuit is based on the cell shown in Fig. 5.16(b). For strong inversion, the generalized translinear principle [Seevinck, 1991] can be used to analyze the circuit. This principle is based on the fact that drain current I_{DS} and Gate-to-Source voltage V_{GS} of a MOS in saturation are related by

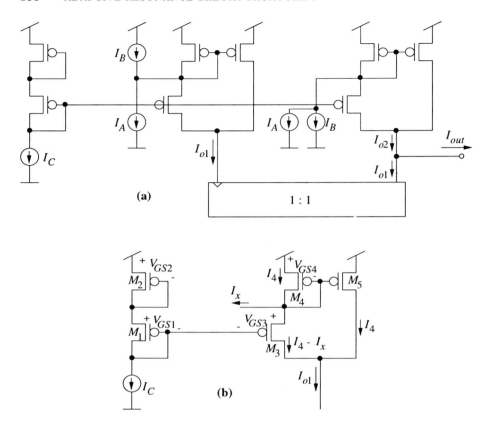

Figure 5.16. (a) Complete Schematic of Current Division Circuit, (b) Basic Processing Circuit

$$V_{GS} = V_T + \sqrt{\frac{I_{DS}}{K}} \tag{5.35}$$

where K is a geometry and technology dependent constant and V_T the MOS threshold voltage. Since for the circuit of Fig. 5.16(b)

$$V_{GS1} + V_{GS2} = V_{GS3} + V_{GS4} \tag{5.36}$$

by applying eq. (5.35) it follows that

$$I_4 = I_C + \frac{I_x^2}{16 I_C} + \frac{I_x}{2} \tag{5.37}$$

Since it must be $I_4 - I_x > 0$ (so that transistor M_3 keeps working in saturation), from eq. (5.37) the condition

$$I_x < 4I_C \tag{5.38}$$

results. Current I_{o1} in Fig. 5.16(b) is given by,

$$I_{o1} = 2I_4 - I_x = 2I_C + \frac{I_x^2}{8I_C} \tag{5.39}$$

Using eq. (5.39) for I_{o1} and I_{o2} in Fig. 5.16(a) results in

$$I_{out} = I_{o2} - I_{o1} = \left(2I_C + \frac{(I_A + I_B)^2}{8I_C}\right) - \left(2I_C + \frac{(I_A - I_B)^2}{8I_C}\right) = \frac{I_A I_B}{2I_C} \tag{5.40}$$

By making I_B to be an internal fixed bias current, output current I_{out} performs the division operation of current I_A over current I_C.

In weak inversion, by applying the classical translinear principle [Gilbert, 1975] to the loop $M_1 - M_2 - M_3 - M_4$ of Fig. 5.16(b) results in

$$I_C^2 = I_4(I_4 - I_x) \tag{5.41}$$

which yields

$$I_{o1} = 2I_4 - I_x = 2I_C \sqrt{1 + \left(\frac{I_x}{2I_C}\right)^2} \tag{5.42}$$

If $2I_C \gg I_x$,

$$I_{o1} \approx 2I_C + \frac{I_x^2}{4I_C} \tag{5.43}$$

Applying this to the circuit of Fig. 5.16(a) reveals that the corresponding output current is

$$I_{out} \approx \left(2I_C + \frac{(I_A + I_B)^2}{4I_C}\right) - \left(2I_C + \frac{(I_A - I_B)^2}{4I_C}\right) = \frac{I_A I_B}{I_C} \tag{5.44}$$

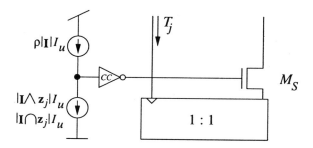

Figure 5.17. Circuit Diagram for Vigilance Circuit

Note that although the circuit of Fig. 5.16 operates both in strong and in weak inversion, it must be biased to operate only in one of the two regions. The reason is that the coefficient of the division is different for both cases, and if the bias conditions are such that the circuit is transitioning from weak to strong inversion (or vice versa) the coefficient is not constant which results in a different operation than the division.

The Vigilance Circuit

Before the computed currents T_j compete in the WTA, the vigilance criterion is checked for each category line j. This is achieved by the circuit shown in Fig. 5.17. Currents $|\mathbf{I} \wedge \mathbf{z}_j|I_u$ or $|\mathbf{I} \cap \mathbf{z}_j|I_u$ are generated and compared against current $\rho|\mathbf{I}|I_u$ by current comparator CC [Rodríguez-Vázquez, 1995] (see Fig. 3.4(a)). Current $|\mathbf{I}|I_u$ is generated by all bottom cells and then mirrored by an adjustable gain current mirror to produce the current $\rho|\mathbf{I}|I_u$, which is then distributed to all *Right* cells. If

$$\rho|\mathbf{I}|I_u \leq |\mathbf{I} \wedge \mathbf{z}_j| \text{ or } |\mathbf{I} \cap \mathbf{z}_j| \tag{5.45}$$

transistor M_S is ON and current T_j competes in the WTA. If not, then transistor M_S is OFF and the effect is as $T_j = 0$.

Winner-Takes-All

The WTA is made of a cascade of two WTA circuits. Each WTA is based on the same circuit described in Section 3.3 [Lazzaro, 1989]. Only one WTA branch is shown in Fig. 5.18. The main transistors for this branch are M_1 and M_2 (for the first WTA circuit), and M_1' and M_2' (for the second WTA circuit). Transistors $M2c$ and $M2c'$ are cascode transistors that improve the gain of each WTA branch. Two WTA circuits are cascaded to improve the discriminatory ability of the circuit. Node *WTAcommon*1 is common for all

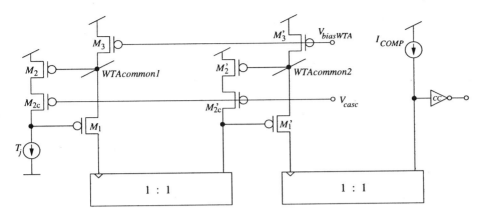

Figure 5.18. One Branch of the WTA Circuit

branches ($j = 1,\ldots M$) of the first WTA circuit, while node $WTAcommon2$ is shared by all branches of the second WTA circuit. To identify the winning branch, the output current is compared against I_{COMP} by current comparator CC [Rodríguez-Vázquez, 1995] (see Fig. 3.4(a)).

Not shown in Fig. 5.18, the circuit includes a shift register for disabling specific WTA branches. This can be used to only allow one uncommitted node to compete in the WTA, and to disable faulty lines j in the chip.

Encoder and Decoder Circuits

The encoder circuit in Fig. 5.11 transforms into a binary word the line number J that has been chosen as the winner. This word can be read from the outside and therefore reveal the category chosen for coding.

Besides this, a decoder circuit is also included to externally force a "fake" winner. This is useful for test purposes or for systematically reading out the whole chip weights template.

RAM Circuit

In order to emulate ARTMAP or Fuzzy-ARTMAP behavior, a simple RAM circuit can be added. This circuit stores, for each ARTa category j its corresponding ARTb class.

Initially the RAM can be set to all bits equal to '1'. Then, during training mode an input pattern **I** is given to the ART1 or Fuzzy-ART system and a category J is selected for storage. If the RAM content for this category is all

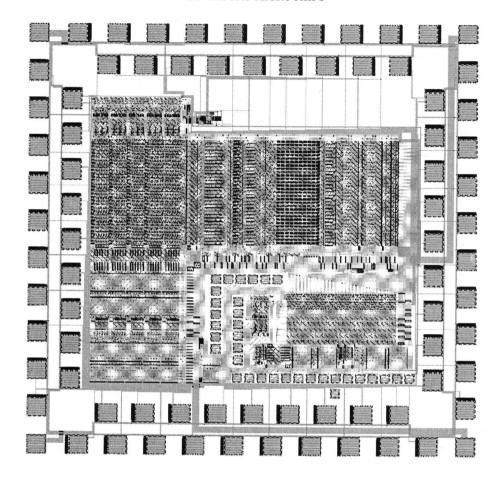

Figure 5.19. Complete Layout of Small Size Prototype Test Chip

'1' then its corresponding class **b** (which is given coded as a digital word) is stored into this RAM. If the content of the RAM is not all '1' and is different than the word for class **b**, then the vigilance parameter of ARTa has to be increased until a new category appears, and so on. During recall mode, after category J is selected, one only needs to read out the RAM content for this category to see the predicted class **b**. If the output is all '1' then the prediction is a "don't know".

5.3 CONCLUDING REMARKS

At the time of this writing a first prototype for this second generation ART chip is under study. It is a reduced (small area) version that implements a Fuzzy-ART synaptic array of 5×31 or an ART1 synaptic array of 25×31 cells. The layout of this chip, which has been developed for a 3-metal, single poly $0.5 \mu m$ CMOS process is shown in Fig. 5.19. Hopefully, depending on funding availability we might be able to report some time in the future on results of a real size prototype.

6 ANALOG LEARNING FUZZY ART CHIPS

MARC COHEN, PAMELA ABSHIRE, JEREMY LUBKIN, AND GERT CAUWENBERGHS

Johns Hopkins University

6.1 INTRODUCTION

This chapter addresses issues of continuous-valued, continuous time implementations of Fuzzy–ART. Inputs to these chips are discrete-time/space, continuous valued signals. We present two analog VLSI chips that implement different variants on Fuzzy ART algorithms. Key design considerations are; on-chip learning, analog weight storage, high integration density and low power dissipation.

The first chip [Cohen, 1998] is a full implementation of analog Fuzzy–ART including complement encoding, fuzzy–min and vigilance. It uses volatile weight storage and counteracts weight leakage by incorporating refresh in the learning rule. It also has a tracking mechanism to automatically reallocate a category when the vigilance condition is not met for any existing category.

The second chip [Lubkin, 1998] achieves long-term dynamic weight storage by quantization and partial incremental refresh [Cauwenberghs, 1992]. It does not explicitly perform complement encoding, nor the creation of new categories from vigilance, in the present implementation. Its unit cell is more compact

($71\lambda \times 71\lambda$) than the first chip's ($93\lambda \times 193\lambda$) and can also be configured as a conventional learning vector quantizer (LVQ) [Gersho, 1992], if so desired.

6.2 SUMMARY OF THE FUZZY–ART ALGORITHM

For a detailed exposition of the algorithm, we refer to [Carpenter, 1991c], and also Chapter 1, where several variants of Fuzzy ART have been presented. Here we focus on the implemented form, and define the equations with our notation used to represent the signals.

We assume an N-dimensional analog input $\mathbf{I}$, and M weight templates $\mathbf{z}_j$ of same dimension, each representing one category. The *choice function* T_j for each category j is computed as

$$T_j = \frac{|\mathbf{I} \wedge \mathbf{z}_j|}{\alpha + |\mathbf{z}_j|} \qquad (6.1)$$

and the *vigilance condition* for each category is formulated as

$$\frac{|\mathbf{I} \wedge \mathbf{z}_j|}{|\mathbf{I}|} > \rho \qquad (6.2)$$

where ρ is the vigilance parameter. Only when a category meets the vigilance condition, it enters competition for the highest T_j, which identifies the winning output category.

Once the winning category has been selected, the weights belonging to that category are updated according to the learning rule:

$$\Delta \mathbf{z}_j = \beta \left(\mathbf{I} \wedge \mathbf{z}_j - \mathbf{z}_j \right) \qquad (6.3)$$

If no category is vigilant to enter the competition, a new category is created, and initialized with *fast learning*, $\mathbf{z}_j \equiv \mathbf{I}$.

Category Selection

In both chip implementations, the vigilance condition is checked before a category enters the competition, eliminating the need to iterate the search until a vigilant winning category is found. The only complication arises in the case where there are no vigilant categories. This case triggers the creation of a new category with *fast learning* which replaces the "least active" among the existing categories, as described in more detail for the first chip implementation below. This is necessary in hardware, as the number of categories is hardwired.

The fuzzy min operator $\wedge$ is defined component-wise as

$$I_i \wedge z_{ij} = \min(I_i, z_{ij}) \ . \qquad (6.4)$$

Inputs to the classifier are complement-encoded, supplying the complements $\overline{I_i} = I_{max} - I_i$ along with all inputs I_i. Explicitly,

$$|\mathbf{I} \wedge \mathbf{z}_j| = \sum (\min(I_i, z_{ij}) + \min(\overline{I_i}, z'_{ij})) \qquad (6.5)$$

where the template weights for the complementary input components z'_{ij} are independent from z_{ij}. The complement-encoding of the inputs provides automatic normalization, $|\mathbf{I}| \equiv N \cdot I_{max}$.

Learning

Expanding the fuzzy min operator, the weight updates (6.3) can be expressed as

$$\Delta z_{ij} = \lambda(I_i, z_{ij})\,(I_i - z_{ij}) \qquad (6.6)$$

where

$$\lambda(I_i, z_{ij}) = \begin{cases} 1 & \text{for initial coding and recoding} \\ \beta & \text{when } I_i \leq z_{ij} \\ 0 & \text{when } I_i > z_{ij} \end{cases} \qquad (6.7)$$

This modulation of the learning rate λ is easier to implement in hardware, since it reduces to directly modulating the current supplying the weight updates.

Note that after coding/recoding the update can only decrease the stored weight value, so it is important that all weights be initialized to their maximum value.

6.3 CURRENT–MODE FUZZY–ART CHIP

For the first chip presented here, we describe our full implementation of the Fuzzy–ART algorithm with a small test–chip having $N = 4$ and $M = 8$ [Cohen, 1998]. We describe modifications to the algorithm which accommodate weight leakage and finite M, the former present in this chip and the latter fixed for any hardware implementation.

Modifications to the Algorithm

As equations (6.3) and (6.7) indicate, no update occurs when $I_i > z_{ij}$. Furthermore, we anticipate that diode leakage currents onto the weight storage capacitors will cause the weights to decay over time. This decay causes the weights to move in the same direction as the normal update rule. To counteract this problem of category drift we propose the following modification to the

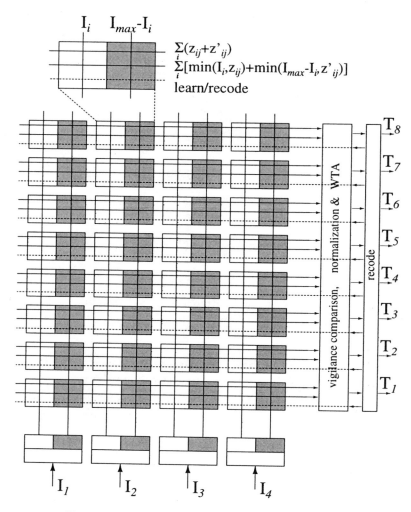

Figure 6.1. Chip architecture for $N = 4$ and $M = 8$

learning rule (6.7):

$$\lambda(I_i, z_{ij}) = \begin{cases} 1 & \text{for initial coding and recoding} \\ \beta & \text{when } I_i \leq z_{ij} \\ \beta_{\text{small}} & \text{when } I_i > z_{ij} \end{cases} \quad (6.8)$$

This modification allows a small update towards the input in the case where the input exceeds the weight. This ensures *refresh* of the drifting weights towards the centroid of the input distribution belonging to that particular category.

This refresh does not significantly disturb the asymmetry of weight update so long as the learning rate β is sufficiently larger than β_{small}.

A second modification, as mentioned already, pertains to the mechanism of assigning new categories. Since M is fixed in the implementation we cannot keep assigning new categories whenever ρ is not exceeded. Instead, it is possible to recode the existing category that has been least frequently chosen in the past. Activity of a category is tracked by a lossy state variable A_j that is updated incrementally whenever that category is selected as the winner. A parallel search for the minimum A_j selects the category to be recoded. Of course, when the category capacity of the chip is not exceeded, existing categories will not be recoded. Recoding can be disabled if desired, for example when it is important to maintain a representation of an infrequently visited category.

Hardware Implementation

We have adopted a modular, scalable architecture. Figure 6.1 depicts a 4-input-8-output network. In general, N-dimensional inputs are presented along the bottom of the network and are complement-encoded into $2N$ signals by the input blocks. Current-mode inputs I_i are applied either directly as currents, or in the form of a gate voltage of an nMOS transistor sourcing the current. The dynamic range of the input signals is determined by the choice of $I_{\max}$ for complement encoding. The complement-encoded inputs I_i and $\overline{I_i}$ are passed to an array of cells each of which computes the choice functions and vigilance conditions, and performs the learning locally for each weight. The shaded cells represent the circuit elements computing with complement encoded signals.

The fuzzy min computation for choice function and vigilance is performed in the current domain so that it is natural to sum the contributions from all cells in that row. The weights z_{ij}, also represented as currents, are also summed along each row. The summed fuzzy min currents are compared with the vigilance ρ using a simple current comparator, and normalized to the summed weights for the choice function computation T_j using a translinear normalizer [Gilbert, 1990b]. If the vigilance criterion is satisfied, the normalized output T_j is passed to a current-mode winner-takes-all circuit [Lazzaro, 1989] which determines the category which best matches the input. A logic "high" signal is broadcast to all the cells along the winning row, and a logic "low" elsewhere, so only the winning row is subject to learning.

In this first chip implementation, inputs are explicitly complement-encoded, and pairs of identical cells are used in the array to perform the computations on I_i and $\overline{I_i}$, and update the weights z_{ij} and z'_{ij}. Since the function of both cells are identical, we will not make the distinction between them in the circuit description below.

Our implementation automatically selects between three different learning rates for each stored weight: a high rate β_{big} for initial coding or recoding, a

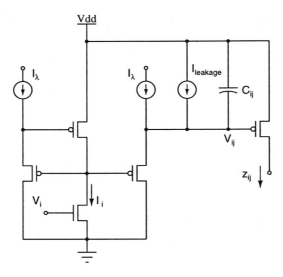

Figure 6.2. The learning cell

moderate rate β for regular learning, and a low rate β_{small} for refresh learning. β_{big} is used only when a row is selected for coding/recoding, and it is applied to the entire row which is being coded/recoded so that the resulting weight values are identical to the input. β and β_{small}, in contrast, are used for incremental updates. When a row is selected as the winner, each cell in that row locally determines whether the input is less than or greater than the stored weight in that cell, and the appropriate learning rate is selected per equation (6.8).

A schematic of the learning and weight storage cell is shown in Figure 6.2. The cell consists of a translinear log domain filter [Himmelbauer, 1996] with a 1pF storage capacitor on the gate of the output pMOS transistor, implementing the weight z_{ij}. The weight current z_{ij} linearly follows the input current I_i, first-order lowpass filtered with a time constant inversely proportional to I_λ. The circuit is operated as a charge-injection-free charge pump by briefly (1μs) pulsing the learning current I_λ [Cauwenberghs, 1992]. The filter operates only during the brief learning pulse which is applied to the winning row.

If ρ is not exceeded for any row, then the least frequently used row is selected for recoding: the "high" logic level signal is broadcast to only that row, and the learning rate for that row is temporarily set to β_{big}.

A small test network comprising 4 inputs and 8 outputs has been fabricated through the MOSIS foundry service in a tiny-chip 1.2μm nwell double-poly technology. Results from a unit cell test-structure are presented here. The unit cell which performs the Fuzzy min and learning operations measures 100 μm by 45 μm.

Fuzzy ART with the modified learning rule

Origins of VLSI Weight Decay. In each cell of the array, the weight current z_{ij} (or z'_{ij}) is sourced by a pMOS transistor (see Figure 6.2) whose gate voltage is set by the storage capacitor. The storage capacitors in our implementation are not completely insulated by gate oxide. Learning is accomplished when current flows to or from the capacitor, and this current is sourced by pMOS transistors. Leakage currents I_{leakage} from these p-type diffusion areas will tend to drive the capacitor voltages toward the positive rail. Thus the voltage on each capacitor C_{ij} will decay linearly from an initial value toward the positive rail (V_{dd}). The voltage decay rate can be estimated from typical 1.2 μm CMOS process parameters as follows:

$$\left| \frac{dV_{ij}}{dt} \right| = \frac{I_{\text{leakage}}}{C_{ij}} \approx \frac{10^{-14}\text{A}}{10^{-12}\text{F}} = 10^{-2}\text{V/s}. \tag{6.9}$$

Since the current is an exponential function of the gate voltage when the transistor is operating in the subthreshold regime, an exponential decay of the currents z_{ij} and v_{ij} will result, with time constant:

$$\tau_{\text{leakage}} = U_T C_{ij}/(\kappa I_{\text{leakage}}) \approx 4.2\text{s} \tag{6.10}$$

where κ is the subthreshold slope factor and U_T is the thermal voltage. If we choose a presentation rate of $F = 10000$ presentations per second, the weight decay constant γ will be (per presentation):

$$\gamma = 1/(\tau_{\text{leakage}} \cdot F) = 2.4 \times 10^{-5} \tag{6.11}$$

Compensation for Weight Decay. During operation of the Fuzzy ART algorithm, each category is coded initially via fast learning and then undergoes a period of normal learning during which the category boundaries are enlarged by successive input presentations. As described by Carpenter and Grossberg, the maximum size of the category box is related to the vigilance parameter ρ. Eventually the categories reach a steady state condition wherein their weights describe the distribution of inputs well enough that no further learning is necessary, and ideally all weights should remain constant thereafter. However, weight decay will have the effect of further enlarging the boxes until the category is so large that the vigilance comparison cannot be satisfied for any input. Weight decay acts in the same direction as the normal learning. To avoid this instability it is necessary to counteract the weight decay, and a suitable mechanism is provided by the modified learning rule.

It can be shown [Himmelbauer, 1996] that the incremental update implemented by the cell in Figure 6.2 has the form of a first-order lowpass filter:

$$\Delta z_{ij} = I_\lambda (I_i - z_{ij}) \tag{6.12}$$

This update applies to both normal and refresh learning, but with different values for the learning rate I_λ as in equation (6.8). Including weight decay, the above equation becomes:

$$\Delta z_{ij} = I_\lambda (I_i - z_{ij}) - \gamma z_{ij} \tag{6.13}$$

Weight decay happens continuously for all weights, whereas refresh learning occurs only during the learning pulse and only for the weights belonging to the winning category. On average, for M active categories, each of which is selected with equal probability, refresh learning occurs only once for every M input presentations. To maintain stable categories, the refresh learning must balance the weight decay:

$$\Delta t \cdot F \cdot I_{\beta \text{small}} \cdot E[I_i - z_{ij}]/N \geq \gamma E[z_{ij}] \tag{6.14}$$

where Δt is the length of the learning pulse, F is the frequency of presentation and E denotes the expectation operator. This constraint is satisfied if $\gamma \cdot N \ll I_{\beta \text{small}} \cdot \Delta t \cdot F$. In addition, the asymmetry of the original weight update (no decay and no refresh) is preserved if $I_\beta \gg I_{\beta \text{small}}$. The constraints on the learning rates can be summarized as follows:

$$I_\beta \gg I_{\beta \text{small}} \gg \gamma \cdot N/(\Delta t \cdot F) \tag{6.15}$$

This is easily satisfied for anticipated values of leakage currents, presentation frequency and learning currents.

Simulations have been performed [see Figure 6.3] to verify that the refresh learning compensates for weight decay and doesn't affect the asymmetry or stability of the original algorithm. When $I_{\beta \text{small}}$ is set correctly, the weights fluctuate about their appropriate values. When $I_{\beta \text{small}}$ is too small, the weights decay until the category is too large for any input to satisfy the vigilance criterion, and a new category is erroneously created. For the results in Figure 6.3, two inputs were each uniformly distributed in [0.49, 0.51] and the vigilance during training was set at 0.8, so that one category described the input distribution. The top plot shows the decay of the weights when $I_{\beta \text{small}}$ is too small. At the given vigilance, a new category will be created when these weights decay to 0.4. The bottom plot shows that the weights fluctuate about the correct value for an appropriate value of $I_{\beta \text{small}}$.

Test Results: Fuzzy Min and Learning

We performed experimental tests on a fabricated prototype to validate performance. The results presented here were obtained from an isolated cell so as to characterize the Fuzzy min computation and learning.

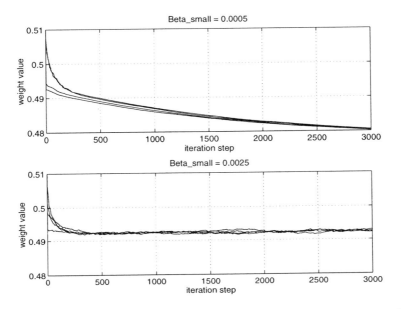

Figure 6.3. Evolution of weights for different β_{small}. $\beta = 0.05$, $\gamma = 2.4 \times 10^{-5}$

To test the Fuzzy min computation we disabled learning and manually set the weight current z_{ij}. We then ramped the input current I_i from a value less than z_{ij} to a value greater than z_{ij}. When the input is less than the stored weight, the output follows the input. Once the input exceeds the stored weight, the output remains constant at the stored weight value. Figure 6.4 plots the measured results for I_i and z_{ij} ranging from about $3nA$ to $1000nA$ clearly demonstrating $\min(I_i, z_{ij})$.

To demonstrate normal learning (I_β), z_{ij} was initially held fixed at a value greater than I_i and then released to adapt down to I_i. This normal learning proceeded at a fast rate (time constant approximately $35\mu sec$). For refresh learning ($I_{\beta \text{small}}$), z_{ij} was initially held fixed at a value less than I_i and then released to adapt up slowly to I_i (time constant approximately $200\mu sec$). Figure 6.5 shows the measured results for an input current of $10nA$ and a range of initial weight values. The decaying traces indicate z_{ij} adapting from $100nA$ ($30nA$) down to $10nA$ with normal learning and the slowly increasing traces indicate z_{ij} adapting from $1nA$ ($3nA$) up to $10nA$ with refresh learning.

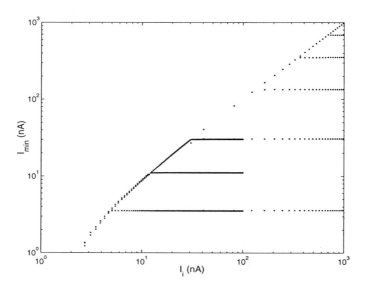

Figure 6.4. Measured Fuzzy-min Computation

6.4 FUZZY–ART/VQ CHIP

The second chip presented here, a 16 × 16 Fuzzy–ART classifier and vector quantizer (VQ), is implemented in current-mode CMOS technology for low-power operation, and integrates learning as well as long-term dynamic capacitive storage of the analog templates using an incremental partial refresh scheme [Cauwenberghs, 1992].

Algorithms and Architecture

The architecture of the hybrid implementation is shown in Figure 6.6. The core contains an array of 16 × 16 template matching cells interconnecting rows of templates with columns of input components. Each cell constructs a distance $d(I_i, z_{ij})$ between one component I_i of the input vector **I** and the corresponding component z_{ij} of one of the template vectors $\mathbf{z}_j$. For Fuzzy ART, this is the Fuzzy min as defined above; for VQ the distance is the absolute difference $|I_i - z_{ij}|$, [Lubkin, 1998]. The component-wise distance is accumulated across inputs along template rows to construct T_j, and presented to a winner-take-all (WTA), which selects the single winner

$$J = \arg\max_j T_j \,. \tag{6.16}$$

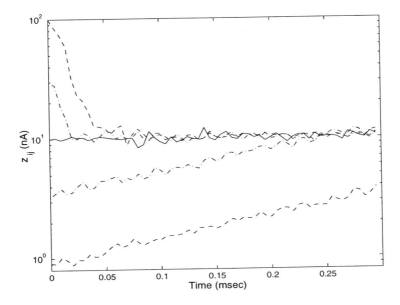

Figure 6.5. Measured Normal and Refresh Learning

In the case of VQ, T_j is constructed by accumulating $d(I_i, z_{ij})$ without weight normalization and vigilance conditioning.

VLSI-Friendly Choice Function. Similar to the modifications outlined in Chapter 2 for ART1, a "VLSI-friendly" version of the Fuzzy ART choice function is used here. It uses the fact that the inputs are complement-encoded. Thus,

$$\begin{aligned}|\mathbf{I} \wedge \mathbf{z}_j| &= \sum_i (\min(I_i, z_{ij}) + \min(\overline{I_i}, z'_{ij})) \\ &= NI_{\max} + \sum_i (\min(I_i, z_{ij}) - \max(I_i, \overline{z'_{ij}}))\end{aligned} \quad (6.17)$$

where $\overline{z'_{ij}} = I_{\max} - z'_{ij}$ is the complement of the weight corresponding to the complement-encoded input. This and similar expressions for $|\mathbf{z}_j|$ allow to

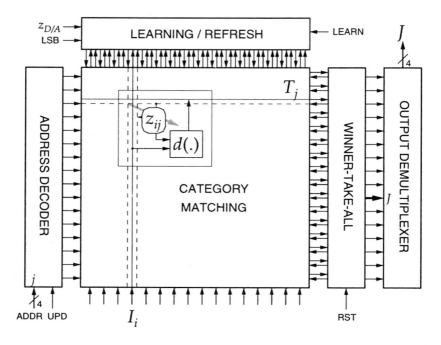

Figure 6.6. Parallel VLSI architecture for Fuzzy ART and VQ, including template learning and refresh functions.

expand the choice function as

$$\begin{aligned} T_j &= \frac{|\mathbf{I} \wedge \mathbf{z}_j|}{\alpha + |\mathbf{z}_j|} \\ &\approx \frac{1 + \frac{\sum_i (\min(I_i, z_{ij}) - \max(I_i, \overline{z'_{ij}}))}{NI_{\max}}}{1 + \frac{\sum_i (z_{ij} - \overline{z'_{ij}})}{NI_{\max}}} \\ &\approx 1 + \frac{\sum_i (\min(I_i, z_{ij}) - \max(I_i, \overline{z'_{ij}}) - z_{ij} + \overline{z'_{ij}})}{NI_{\max}} \end{aligned} \qquad (6.18)$$

assuming $NI_{\max}$ is much larger than other terms in the expression. Even in cases where this assumption breaks down, the approximation still preserves monotonicity in the choice function. The simplification is similar to the VLSI-friendly version of the ART1 choice function in Chapter 2, except that α has been suppressed. As before, this simplification eliminates the need to divide analog signals.

We further simplify the algorithm, and eliminate the need of computing and subtracting the summed weights $|\mathbf{z}_j|$ altogether. Since at initial coding $z_{ij} \equiv \overline{z'_{ij}}$, and since the weights z_{ij} and z'_{ij} only decrease under the learning updates, it follows that $z_{ij} \leq \overline{z'_{ij}}$. Thus,

$$\min(I_i, z_{ij}) - \max(I_i, \overline{z'_{ij}}) - z_{ij} + \overline{z'_{ij}} = \min(I_i, \overline{z'_{ij}}) - \max(I_i, z_{ij}) \quad (6.19)$$

and the "VLSI-friendly" choice function becomes even friendlier:

$$\begin{aligned}
T_j &= \frac{NI_{\max} - \sum_i(\max(I_i, z_{ij}) - \min(I_i, \overline{z'_{ij}}))}{NI_{\max}} \\
&= \frac{2NI_{\max} - \sum_i(\max(I_i, z_{ij}) + \max(\overline{I_i}, z'_{ij}))}{NI_{\max}} \\
&= 2 - \frac{|\mathbf{I} \vee \mathbf{z}_j|}{NI_{\max}}. \quad (6.20)
\end{aligned}$$

In other words, maximizing the fuzzy min choice function normalized by the weights is, approximately, equivalent to minimizing a modified fuzzy max choice function, without weight normalization.

It can be verified that this new choice function for the Fuzzy-ART algorithm produces valid clustering behavior[1]. However, for correct Fuzzy-ARTMAP operation, it is not possible to neglect α in the approximation of the choice function, at the risk of invalidating certain properties such as "Direct Access to Subset and Superset Patterns". Nevertheless, a non-zero value for α is easily accounted for in the above approximations, changing (6.20) into

$$\begin{aligned}
T_j &\approx \frac{1 + \frac{\sum_i(\min(I_i, z_{ij}) - \max(I_i, \overline{z'_{ij}}))}{NI_{\max}}}{1 + \frac{\sum_i(z_{ij} - \overline{z'_{ij}})}{NI_{\max} + \alpha}} \\
&\approx 1 + \frac{\sum_i(\min(I_i, z_{ij}) - \max(I_i, \overline{z'_{ij}}) - (1-\alpha')(z_{ij} + \overline{z'_{ij}}))}{NI_{\max}} \quad (6.21) \\
&= 2 - \alpha' - \frac{(1-\alpha')|\mathbf{I} \vee \mathbf{z}_j| - \alpha'|\mathbf{I} \wedge \mathbf{z}_j|}{NI_{\max}} \quad (6.22)
\end{aligned}$$

where $\alpha' = \alpha / NI_{\max}$. Thus, for Fuzzy ARTMAP with nonzero value for α, a valid choice function can still be constructed by linearly combining fuzzy max and min operations.

The chip computes, for each template row, both fuzzy max and fuzzy min distances between input and template in parallel:

$$\begin{aligned}
T_j^+ &= |\mathbf{I} \vee \mathbf{z}_j| \\
T_j^- &= |\mathbf{I} \wedge \mathbf{z}_j| \quad (6.23)
\end{aligned}$$

where T_j^- is used for the vigilance condition, and T_j^+ (or $(1-\alpha')T_j^+ - \alpha' T_j^-$) is used for the choice function, which enters competition for the minimum value

---

[1] For an empirical verification, just substitute the computation of the choice function in the file "fuzzy.m" of Appendix A by `T(j)=2-norm(max(a,wij(j,:)),1)/(2*np);`

when the vigilance condition is met. When the chip is configured in VQ mode, the signals $T_j{}^+$ and $T_j{}^-$ are combined differentially to construct the mean absolute difference (MAD) distance [Cauwenberghs, 1995b]

$$T_j{}^+ - T_j{}^- = |\mathbf{I} - \mathbf{z}_j| \ . \tag{6.24}$$

Learning and Refresh. Learning is performed by selecting the winning template J and producing an incremental update $\Delta \mathbf{z}_J$ in the stored vector $\mathbf{z}_J$ towards the input vector, according to a modified version of (6.6). The learning rate λ is modulated according to (6.7), except in VQ operation for which the learning rate is constant, $\lambda \equiv \beta$. In the case of Kohonen self-organizing maps [Hochet, 1991], the neighbors of the winner, $j = J \pm 1$, are also updated according to (6.6) to preserve topological ordering in the digital coding.

The modification in the update rule (6.6) is to fix the update amplitude by thresholding:

$$\Delta z_{iJ} = \lambda(I_i, z_{iJ}) \operatorname{sgn}(I_i - z_{iJ}) \ . \tag{6.25}$$

A constant-amplitude, variable-polarity discrete update is easier to implement than a continuous update, and gives superior results in the presence of analog imprecisions in the implementation [Cauwenberghs, 1997]. The granular effect of coarse updates is avoided by reducing the update constant β.

Dynamic refresh for long-term analog storage of the weights z_{ij} is achieved using the technique of binary quantization and partial incremental refresh [Cauwenberghs, 1992]. The technique counteracts drift due to leakage in volatile storage, by maintaining the analog value near one of quantized levels. The stable levels of the dynamic memory are defined by the transition levels of a binary quantization function Q, which maps analog values to binary values $\{-1, 1\}$. When $Q(z_{ij}) = 1$, the analog memory value z_{ij} is slightly decreased, and conversely when $Q(z_{ij}) = -1$, z_{ij} is slightly increased:

$$\Delta z_{ij} = -\mu \, Q(z_{ij}). \tag{6.26}$$

Periodic iteration of updates (6.26) yields long-term stable memory as long as the update amplitude μ is larger than the worst-case drift in between refresh iterations, and significantly smaller than the separation between memory levels [Cauwenberghs, 1992].

VLSI Implementation

The circuits are implemented in current-mode CMOS technology, with MOS transistors operated in subthreshold for low-power dissipation. The use of lateral bipolar transistors offers the advantages of a BiCMOS process while maintaining full compatibility with standard (digital) single-poly CMOS processes.

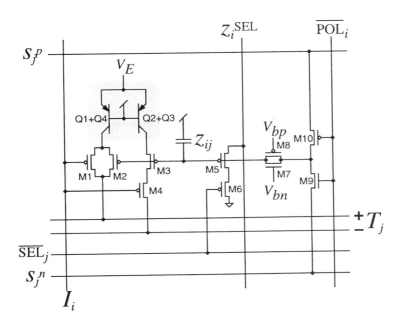

Figure 6.7. Circuit schematic of the Fuzzy ART and VQ template matching cell, with integrated learning and template refresh. The dashed inset indicates a matched double pair of lateral bipolar transistors.

Distance Estimation. The circuit diagram of the distance estimation cell is shown in Figure 6.7, and the layout of the cell, measuring 71 μm × 71 μm in 2 μm CMOS technology, is given in Figure 6.8.

As explained above, the choice function and vigilance measures for Fuzzy ART, as well as the distance metric for VQ, are decomposed in terms of the fuzzy maximum and the minimum of I_i and z_{ij}, accumulated separately onto two wires T_j^+ and T_j^- (6.23) and combined outside of the array. Their computation is performed by modulating the Early effect (collector conductance) of a matched (double) pair of bipolar transistors Q1-Q4 and Q2-Q3, by means of MOS transistors M1, M2, M3 and M4 connected as source followers. Parallel and series connections in the source followers yield the maximum and minimum of I_i and z_{ij}, respectively, in the output currents.

A centroid geometry, shown in Figure 6.9, is used for improved matching between the bipolar transistor currents that supply the differential output. By combining collector outputs in pairs $C_1 + C_4$ and $C_2 + C_3$, systematic variations and gradients in geometry are cancelled to first order. Matching is important, since the collector conductance is relatively small. The Early effect

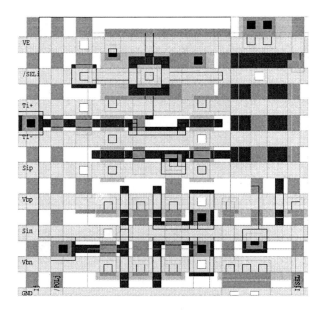

Figure 6.8. Layout of the Fuzzy ART and VQ template matching cell, of size 71 μm × 71 μm in 2 μm CMOS technology.

is maximized by using a minimum length geometry for the base, equaling the minimum length of an MOS transistor.

The winner-take-all (WTA) is implemented as a variation on the standard current-mode design in [Lazzaro, 1989; Andreou, 1991] with the addition of a triggered voltage-mode output stage for improved settling accuracy and speed [Cauwenberghs, 1995b].

Learning and Refresh. Transistors M5 through M10 implement the incremental update Δz_{ij}, when the cell is selected either for refresh, or for learning (when $k^{\text{WTA}} \equiv j$). The polarity $\overline{\text{POL}_i}$ of the fixed-amplitude update Δz_{ij} is precisely implemented by means of a binary controlled charge pump [Cauwenberghs, 1992]. The charge pump is free of switch charge injection parasitics, by avoiding clock signals on the MOS gates that couple capacitively into the storage capacitor. V_{bn} and V_{bp} are biased deep in subthreshold for precisely controlled increments and decrements as small as 10 μV. The timing of the update (and selection of the template) is performed by means of signals S_n^j and S_p^j [Cauwenberghs, 1997].

The update polarity is computed externally to the array, by circuitry common for all cells on the same column, shown at the top of Figure 6.6. This

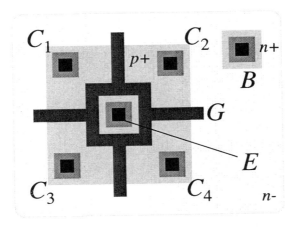

Figure 6.9. Centroid geometry of the matched double pair of lateral bipolar transistors, in conventional n-well CMOS technology.

arrangement is most space efficient since only one row of cells needs to be updated at once. A global signal (LEARN) selects the mode of operation, learning or refresh.

Figure 6.10 shows the simplified schematic of the external learning cell, one per column of VQ distance cells. The circuit receives the selected analog template value z_{ij} on the line Z_j^{SEL} along with the input I_i which are used to generate the update polarity $\overline{POL}_i$ and supply it to the selected distance cell. When a distance cell is selected, switch M6 is closed and transistors M5-M6 along with M11-M13 implement a comparator. The update is performed in the cell according to $\overline{POL}_i$ by activating the signals S_n^j and S_p^j for the entire row of selected cells.

The selection of $\lambda(I_i, z_{ij})$ in (6.7) is implicit in the computed polarity of the update in (6.25), and is automatically enforced by limiting updates to negative polarities only. This is implemented by pulling V_{bn} in Figure 6.7 to zero whenever LEARN is active, in Fuzzy ART mode. The remaining V_{bp} pMOS tail current then provides the negative β updates.[2]

In learning mode (LEARN≡1), the polarity $\overline{POL}_i$ is computed by comparing z_{iJ} with the input I_i, yielding a fixed-size update according to (6.25). In refresh mode (LEARN≡0), z_{iJ} is compared with an external reference signal $z_{D/A}$ to construct the binary quantization function used for partial incremen-

---

[2]Note that negative z_{ij} updates correspond to positive charge injection on the storage capacitor, since the output currents on the $T_j{}^+$ and $T_j{}^-$ lines decrease when the stored voltage increases.

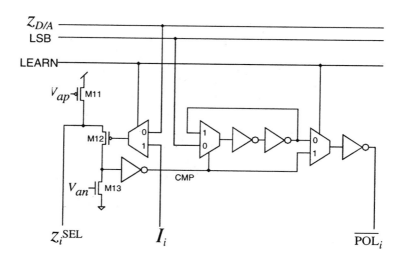

Figure 6.10. Simplified schematic of the learning and refresh circuitry, in common for a column of Fuzzy ART/VQ cells. Analog multiplexers are implemented with complementary CMOS switches.

tal refresh [Cauwenberghs, 1992] in (6.26). As in [Cauwenberghs, 1995a], the binary quantization Q of z_{iJ} is obtained by retaining the least significant bit (LSB) of analog-to-digital (A/D) conversion of z_{iJ}. A single-slope sequential A/D is implemented for simplicity, using a D/A signal on $z_{D/A}$, ramped up in discrete steps synchronously with the alternating LSB. When the comparator flips sign, the instantaneous LSB value is sampled and latched to generate the update polarity $Q(z_{iJ})$, producing an update according to (6.26).

Experimental Results

A micrograph of the 16 × 16 learning Fuzzy ART classifier and vector quantizer, implemented in 2 μm CMOS technology, is shown in Figure 6.11. The chip is fully functional, and dissipates 2 mW from a 5 V supply at a 10 ksample/s parallel data rate.

Experimental results in VQ mode are described in [Lubkin, 1998]. The results reported here are limited to Fuzzy ART operation.

The experimental fuzzy min distance metric is illustrated in Figure 6.12, obtained by sweeping one of the 16 inputs while fixing the other inputs to the template values.

Results for dynamic refresh of the templates at 32-level quantization are shown in Figure 6.13, obtained by observing the drift in stored voltage level on

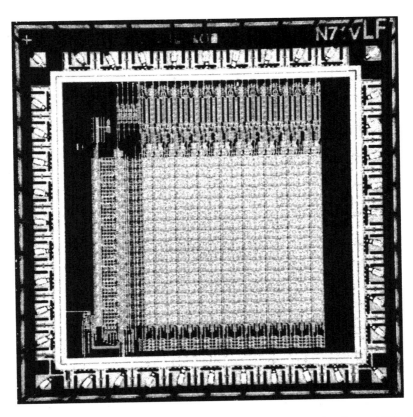

Figure 6.11. Chip micrograph of the 16 × 16 array, analog learning Fuzzy ART classifier and VQ. The die size is 2.2×2.25 mm^2 in 2 μm CMOS technology.

one of the cells over 1000 refresh cycles, for 300 different initial values of the voltage z_{ij}. The corresponding drift without refresh would have been several volts. Refresh tests across the entire array demonstrated long-term storage with minor level drifts over more than 1000 cycles. Further improvements in resolution and stability can be achieved, at some expense in silicon area, by using the A/D/A quantizer in [Cauwenberghs, 1995a].

Learning tests which validate the functionality of the learning Fuzzy ART classifer with adaptive updates according to (6.25) are illustrated in Figure 6.14. The asymmetry in charging rate for upward and downward transitions is a feature of Fuzzy ART. Further experiments are currently underway to characterize system-level performance, for applications in speech coding and reconstruction.

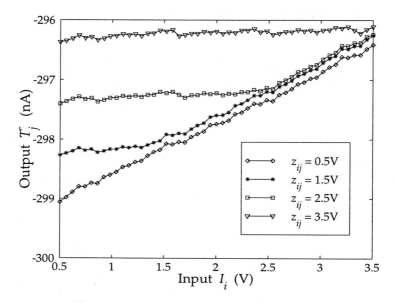

Figure 6.12. Measured distance output for one row of cells, sweeping one input component while fixing the other 15 inputs.

6.5 CONCLUSIONS

We have implemented an asynchronous mixed-mode CMOS VLSI system capable of classifying and learning in real-time. We included in our implementation an addition to the learning rule that gives us the capability to refresh the analog weights and a mechanism for recoding categories which are infrequently used. Both of these new features can be disabled if required. We have verified by simulation that our modified learning rule preserves both the asymmetry and the stability of the original rule in the presence of weight decay in an analog implementation. We have also experimentally verified both the Fuzzy-min computation and learning capability of the fabricated unit cell.

We also presented a parallel architecture and corresponding analog VLSI BiCMOS implementation of a fuzzy ART classifier, fabricated in a single-poly CMOS process using lateral bipolar transistors. The chip is fully functional, and can be configured for variants on fuzzy ART such as VQ and Kohonen self-organizing maps. The chip incorporates analog storage of the templates, sharing the same circuitry used for learning. With a dense cell size of 71×71 λ units in scalable MOSIS technology, the integration of a 256-input, 1024-category classifier is feasible on a 1 cm^2 die in a 0.35 μm CMOS process ($\lambda =$

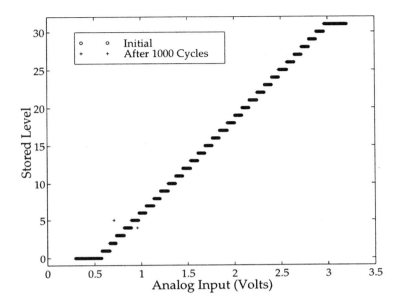

Figure 6.13. Stability of the analog memory array. Drift over 1000 self-refresh cycles, for 300 different initial conditions.

0.2μm). Inclusion of Fuzzy ARTMAP capability would reduce the density roughly by a factor two, still supporting 256 inputs and 512 categories.

Acknowledgements This work was supported by a Multi-University Research Initiative between Boston University and Johns Hopkins University (ARPA and ONR Grant # N00014-95-1-0409), and NSF Career Award MIP-MIP-9702346. Chip fabrication was provided by MOSIS. Pamela Abshire was supported by an NSF Graduate Fellowship.

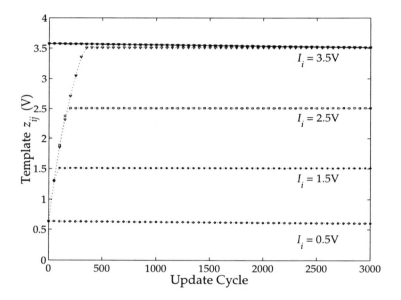

Figure 6.14. Template adaptation of Fuzzy ART in learning mode, under fixed inputs, from low and high template initial conditions.

7 SOME POTENTIAL APPLICATIONS FOR ART MICROCHIPS

When the reader gets to this point of the present book, the strongest criticism he or she might state is *"are there applications for ART based systems where the use of a special purpose ART chip might be justified?"* In this Chapter we intend to provide some hints that might help to answer this question.

If one searches in the specialized literature for ART system applications, they are always implemented using conventional digital processors: either a standard computer, or a special purpose hardware based on a commercial microprocessor. Now the question is, of these applications which would benefit if the ART portion of the system is implemented using a dedicated ART chip, or is incorporated into a larger VLSI circuit? In our opinion, there might be two motivations that could reveal ART microchips to be competitive with respect to standard digital processor based ART implementations:

- **Speed:** when implementing a clustering algorithm on a conventional digital processor it takes a significant amount of clock cycles to perform all the operations required by the algorithm. On the other hand, for the prototypes we presented in Chapter 3 all operations required per input pattern presentation (computation of choice functions, checking of vigilance criterion, selection of maximum, and weights update for winning category) could be performed in less than $2\mu s$, for a prototype chip fabricated in a low performance CMOS process. For applications in which clustering should be done for time frames within the order of microseconds or less, the choice of a dedicated chip (or piece of VLSI within a larger chip) might be necessary. In these cases the ART microchip design should be optimized for maximum speed response.

- **Power Consumption:** for many of the applications, clustering speed is usually not very critical. However, there are many ART-based applications intended for portable (and autonomous) systems. In these cases, the use of conventional digital processors might result non-viable because of power consumption considerations. If a dedicated ART chip design is optimized for power consumption, it might benefit the development of portable systems that require some kind of clustering/classification processing.

There are numerous applications for ART systems reported in the open literature [Carpenter, 1997]. Some of them are:

1. Determination of concentrations of biological substances in tissues [Ham, 1996a]

2. Classification of cardiac arrhythmias [Ham, 1996b]

3. Vehicle interior monitoring for auto alarms [Grasmann, 1997]

4. Boeing parts design retrieval system [Caudell, 1994b]

5. Satellite remote sensing [Baraldi, 1995], [Gopal, 1994]

6. Robot sensory motor control [Bachelder, 1993], [Baloch, 1991], [Dubrawski, 1994], [Srinivasa, 1996]

7. Robot navigation [Racz, 1995]

8. Machine vision [Caudell, 1994a]

9. 3D object recognition [Seibert, 1992]

10. Face recognition [Seibert, 1993]

11. Automatic target recognition [Bernardon, 1995], [Koch, 1995], [Waxman, 1995]

12. Medical imaging [Soliz, 1996]

13. Prediction of protein secondary structure [Mehta, 1993]

14. Strength prediction for concrete mixes [Kasperkiewicz, 1995]

15. Signature verification [Murshed, 1995]

16. Tool failure monitoring [Ly, 1994], [Tarng, 1994], [Tse, 1996]

17. Chemical analysis from UV and IR spectra [Wienke, 1994]

18. Digital circuit design [Kalkunte, 1992]

19. Frequency selective surface design for electromagnetic system devices [Christodoulou, 1995]

20. Chinese character recognition [Gan, 1992], [Kim, 1995]

21. Analysis of musical scores [Gjerdingen, 1990]

Of these we have picked up three examples to describe in this Chapter, because they represent systems in which the use of a dedicated chip (or piece of VLSI inside a larger chip) might result beneficial.

7.1 PORTABLE NON-INVASIVE DEVICE FOR DETERMINATION OF CONCENTRATIONS OF BIOLOGICAL SUBSTANCES

This device is described by Ham and Cohen in a patent document [Ham, 1996a]. It can be used to monitor glucose concentration in diabetic people and control an insulin injector, working in a continuous manner. The device is composed of two parts:

- A physical part which includes a laser diode, optical lenses, mirrors, beam splitters, holographic filters, and wavelength dispersion devices.

- An electronic part including a charge couple device (CCD) image sensor and subsequent circuitry to read the sensed image and process it.

Let us describe each part independently.

Physical Part

A simplified representation of the physical part of this device is shown in Fig. 7.1. It is based on the fact that when a monochromatic light of wavelength λ_o impinges on a tissue sample (like for example, an ear lobe or a finger) Raman scattering is produced. Raman scattering is such that the dispersed light coming out of the illuminated tissue contains wavelengths λ_i whose values are shifted with respect to the incident light wavelength. The set of dispersed light beams with shifted wavelengths and their intensities represent a "fingerprint" of the biological substances and their concentrations in the exposed tissue.

The device described by Ham and Cohen is intended for determining anhydrous D-glucose ($C_6H_{12}O_6$) concentrations. Therefore, they have trimmed the physical part of the device to look at the Raman spectral lines characteristics for this substance. Glucose has a rich Raman spectrum with eight fundamental wavelengths. Table 7.1 shows the eight Raman wavelengths produced by this substance if the incident light wavelength is $\lambda_o = 780nm$.

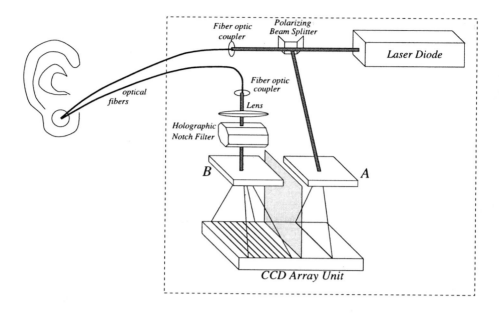

Figure 7.1. Schematic Representation of Physical Part

Table 7.1. Raman Scattered Wavelengths for Glucose with $\lambda_0 = 780nm$

i	λ_i , nm
1	880.2
2	871.4
3	854.7
4	851.3
5	839.9
6	834.8
7	814.4
8	805.5

In Fig. 7.1 a Laser diode generates a monochromatic light beam at $\lambda_o = 780nm$. This beam is divided into two by a polarizing beam splitter. One of the resulting beams is directed to a wavelength dispersion device (A in Fig. 7.1) which deviates the beam according to its wavelength. The resulting dispersed light is sensed by the right half of the CCD array unit shown in Fig. 7.1. This is done for two reasons: (a) to monitor wavelength changes in the light source, and (b) to monitor the intensity of the incident light beam. Wavelength changes are monitored by detecting the position of the incident light on the right half of the CCD array, while intensity is monitored by adding all pixel intensities of the right half CCD array.

The second beam going out of the polarizing beam splitter is directed (for example, through an optical fiber as shown in Fig. 7.1) to the skin tissue. Another optical fiber collects the light scattered by the tissue and directs it to a wavelength dispersion device (B in Fig. 7.1) after lens focusing and holographic notch filtering. The notch filter eliminates all light whose wavelength is equal to that of the laser source light. The light coming out of the wavelength dispersion device B is projected onto the left half of the CCD array unit. On this side of the CCD the intensity as a function of position (i.e., as a function of wavelength) is measured. The outputs of both halves of the CCD array unit are then processed by the electronic part of the system.

Electronic Part

The electronic part of the system performs basically the following processing:

1. data preprocessing for feeding an artificial neural network discriminator, which is a Fuzzy-ARTMAP system,

2. Fuzzy-ARTMAP processing,

3. and taking a decision according to the result of the Fuzzy-ARTMAP output.

The output of the left half of the CCD array provides a spectrum like response $s(\lambda)$. For this signal attention is focused on the 8 spectral lines characteristics of glucose. Variation in position of these lines is known from the output of the right half of the CCD array. For each spectral line m data points are selected (m is an odd number and includes the central point of a spectral line lobe and $(m-1)/2$ points on each side of the lobe). For each of these $8m$ data points, the ratio is computed with respect to the intensity incident on the right half of the CCD array. The whole data acquisition process is repeated q times and results are averaged for noise reduction. The averaged $8m$ samples are complement coded and fed to the Fuzzy-ARTMAP system.

The Fuzzy-ARTMAP system has been trained off line with data that takes into account the following conditions:

1. overlapping spectra of non-interacting biological analytes of varying amounts with the glucose spectrum,

2. spectra associated with molecular interactions of certain biological substances of varying amounts with the glucose spectrum,

3. interactive spectra of glucose and those materials used in the instrument that come into contact with body fluids or tissue to be analyzed,

4. noise and non-linearities associated with the spectroscopic instrument,

5. disturbances due to use of the instrument (like positioning of the tissue with respect to the instrument front end),

6. non-linearities due to the optical properties of skin and/or tissue.

During field operation of the device the Fuzzy-ARTMAP system will respond with either a concentration or a "don't know" answer. If the answer is a concentration, this information can be used to decide whether or not to inject insulin, or to continue monitoring to collect more information. If the answer is a "don't know" then more readings should be performed. If the "don't know" response continues repeatedly then it can be either: (1) the readings correspond to an outlier in which case data is saved for possible use in further retraining, or (2) there is a malfunctioning in the instrument in which case it should be shut down.

In this Fuzzy-ARTMAP application example, the electronic part requires (according to the authors [Ham, 1996a]) a microprocessor because of the extensive processing required by the Fuzzy-ARTMAP algorithm. If the Fuzzy-ARTMAP portion is implemented using a dedicated piece of VLSI the rest of the electronic part does not really need a microprocessor and could be performed with dedicated and compact digital circuitry. In principle, it should be possible to put all the electronics on a single chip. Furthermore, even the CCD array can be substituted by a CMOS image sensor [Andreou, 1995] sharing the same chip with the rest of the electronic circuitry.

7.2 CARDIAC ARRHYTHMIA CLASSIFIER FOR IMPLANTABLE PACEMAKER

An application where an analog low-power ART based neural network might be practical is in implantable pacemakers for efficiently classifying cardiac arrhythmias. As any battery powered implantable device, pacemakers demand very low power consumption, since patients using them have to undergo surgery every time the battery needs replacement. The most sophisticated pacemakers take a decision on whether or not to artificially stimulate the heart muscles based on the shape of recorded electrical impulses. This is a fairly complicated task and can be solved using computationally intensive algorithms implemented on microprocessors [Lin, 1988]. However, this does not help much to reduce the overall system power consumption. Recent work [Ham, 1996b] has shown promising results for efficient cardiac arrhythmia classification based on the use of the Fuzzy-ARTMAP algorithm. Using the circuit implementation ideas presented in this book it can be possible to develop a VLSI system that includes a very low power Fuzzy-ARTMAP for use in an implantable pacemaker. Some extra circuitry would be required to perform the appropriate preprocessing of electrocardiogram signals. Such preprocessing is similar (but simpler) to pre-

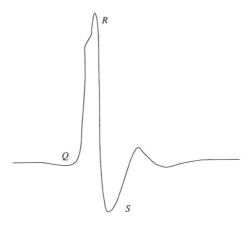

Figure 7.2. Morphology of a QRS complex

processing techniques used for speech recognition, for which some low power analog techniques have been proposed [Kumar, 1997].

An electrocardiogram (ECG) signal is composed of a sequence of QRS complexes. A typical QRS complex is shown in Fig. 7.2. Letters 'Q', 'R', and 'S' are the conventional labels used to denote the different points of the blood conduction cycle in the heart.

A cardiac arrhythmia classification system needs to isolate QRS complexes and, according to their shapes, identify whether or not an arrhythmia is actually taking place. For the Fuzzy-ARTMAP system proposed by Ham and Han [Ham, 1996b], the preprocessing steps are:

1. The ECG signal is bandpass filtered using an analog filter with a passband from 0.1 to $100Hz$. This filter is intended to remove noise caused by respiration, muscle tremors, and spikes.

2. The QRS complexes are extracted based on peak detection techniques to detect the R wave peak.

3. Each extracted QRS segment is centered and width normalized to reduce amplitude variations and Hamming windowed to reduce discontinuities.

4. Linear Predictive Coding (LPC) coefficients are obtained for the normalized QRS segment. LPC coefficients computation is a type of feature extraction for continuous signals, widely used in speech recognition [Rabiner, 1993]. It is based on the idea of trying to predict a signal sample $y(n)$ using older samples. If $\hat{y}(n)$ is the prediction, then

$$\hat{y}(n) = \sum_{i=1}^{p} a_i y(n-i) \tag{7.1}$$

where a_i is the *i-th* LPC coefficient and p the order of the predictor. For each QRS segment, samples $y(n)$ are used to obtain the optimum set of LPC coefficients $\{a_i\}$. Apparently, only a second order predictor is necessary for this application. Therefore, only two LPC coefficients $\{a_1, a_2\}$ are extracted. Ham and Han also computed the mean square value for each QRS segment $\overline{R^2}$. Thus, these sets of three feature parameters $\{a_1, a_2, \overline{R^2}\}$ for each QRS complex are used as inputs to the Fuzzy-ARTMAP network.

5. Finally, the previous three feature parameters are normalized to lie within the [0,1] interval and are complement encoded for proper processing by the Fuzzy-ARTMAP system.

Obviously, all this preprocessing, performed by Ham and Han using conventional computers, might result computationally too expensive to be implemented directly on a VLSI chip for an implantable device. However, these preprocessing techniques are a simplified version of some used for speech recognition [Rabiner, 1993], and for which alternative techniques suitable for low power portable devices have been developed [Kumar, 1997].

Once the extracted features are completely pre-processed, an off-line trained Fuzzy-ARTMAP network takes them as inputs and decides whether or not an arrhythmia is taking place.

7.3 VEHICLE INTERIOR MONITORING DEVICE FOR AUTO ALARM

In a recent patent U. Grasmann [Grasmann, 1997] describes an auto alarm system able to monitor a vehicle's interior and detect when a glass is broken and/or an incursion made. The alarm is based on processing of echoed sound waves and subsequent discrimination which can be done using an ART based neural network.

Fig. 7.3 shows a conceptual block diagram of the complete system. A small controller guarantees proper sequencing of all operations. First, a sound generation transducer SGT is activated through a driver circuit, and a train of sound wave pulses are generated in the vehicle's interior. In the vehicle's interior there is also a collection of sound pick-up units $PU_1, PU_2, \ldots, PU_N$ whose outputs are amplified and time multiplexed to form an input stream signal. The envelope of this signal is extracted using a rectifier and a low-pass filter. Such envelope is then digitized by an A/D converter and the stream stored into a shift register. The length of the stream, which is equal to N times the length of the sequence of pulses generated by the sound generation transducer (SGT), does not have to be too large. The data in the shift register is then fed to an

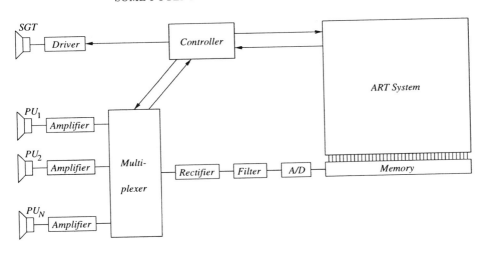

Figure 7.3. Block Diagram for ART based Auto Alarm

off-line trained ART system to discriminate if there is an incursion or window breakage.

The ART system, which according to our opinion should preferably be a Fuzzy-ARTMAP system, should be trained off-line under a variety of incursion situations, window absences, and normal (non-alarm) conditions. Furthermore, every time the alarm is activated when people leave the car, a recording should be taken and added to the ART system database as a normal (non-alarm) situation. This would help to avoid false alarm detection since the vehicle interior might change depending on the objects left in it. The alarm system can operate either in continuous mode or be triggered if there is a sudden noise inside the vehicle (for example, a glass breakage).

In this ART system application, power consumption is not extremely critical. However, this application is a clear consumer market example, where a large number of units might be sold. Under these circumstances it might help to bring costs down by putting all the processing and control electronics in the same VLSI chip. In a standard implementation one would require the use of a commercial microprocessor to emulate the ART system. However, by using the circuit implementation ideas exposed in this book a dedicated ART system could share the same VLSI substrate with the rest of the electronics, resulting in a compact one chip system realization for the complete alarm.

7.4 CONCLUDING REMARKS

As can be seen there is some potential room for Adaptive Resonance Theory Microchips applications. In these days where microchip technology continues shrinking, more and more complex systems can be put together on a single chip (or set of chips). It is possible to add some degree of intelligent processing to a wide variety of products. Many times this intelligent processing can be equivalent to some kind of pattern clustering or classification task. If this is so, there is plenty of room for the ART based clustering and classifying VLSI engines described throughout this book. In this short Chapter we have tried to illustrate this with some application examples. We hope that this book helps to trigger the proliferation of practical systems that include some kind of ART microchip for their internal intelligent processing.

Appendix A
MATLAB Codes for Adaptive Resonance Theory Algorithms

In this Appendix we include some simple "*home-made*" MATLAB codes which help to illustrate and understand the Adaptive Resonance Theory algorithms described in Chapter 1. There are four programs (ART1, ARTMAP, Fuzzy-ART, and Fuzzy-ARTMAP) each of which applies one of the algorithms to a specific problem. The codes are included in this Appendix but can be retrieved also from http://www.imse.cnm.es/~bernabe

A.1   MATLAB CODE EXAMPLE FOR ART1

The *art1* routine generates a sequence of np input patterns of length $n1 \times n2$ and clusters them using the ρ and L values provided. From the MATLAB prompt this routine is called using the command

```
>> art1(n1,n2,rho,L,np)
```

The program then displays in a graphic window the present and previous status of *Input Pattern* (on the left side) and weight templates z_j (on the right side). Patterns are drawn as rectangular boxes of $n1 \times n2$ pixels. The weight template that has been chosen for update (z_J) is drawn surrounded by a yellow line. For example, the command

```
>> art1(5,5,0.4,5,10)
```

would generate a similar sequence than the one shown in Fig. 1.3 of Chapter 1. The program stops after each pattern presentation and corresponding learning, and waits for the user to hit any key before presenting the next pattern. Input patterns are sequentially and iteratively provided until there is no weight update for a complete iteration of input patterns presentation. The program uses the auxiliary functions "*draw_status*" and "*art1plot*" given at the end of this Appendix. The MATLAB code for this program is:

File: "art1.m"

```
function art1(n1,n2,rho,L,np)
clf
n=n1*n2;
z=ones(n,1);
zold=z;
zprev=z;
Jold=1;
Mold=1;
M=1;
first=1;
I='';
for pattern=1:np
  I=[I,round(rand(n,1))];
end
iter=0;
learning=1;
while learning==1
  iter=iter+1;
  for pattern=1:np
    ok=0;
    for j=1:M
      T(j)=norm(min(I(:,pattern),z(:,j)),1)/(L-1+norm(z(:,j),1));
    end
    while ok==0
      [maxT,J]=max(T);
      if rho*norm(I(:,pattern),1) <= norm(min(I(:,pattern),z(:,J)),1)
        ok=1;
      else
        T(J) = -1;
      end
    end
    z(:,J)=min(I(:,pattern),z(:,J));
    if J==M
      M=M+1;
      z=[z ones(n,1)];
    end
    draw_status(M,Mold,pattern,np,J,Jold,I,z,zprev,n1,n2,first,iter)
    first=0;
    Jold=J;
    Mold=M;
    zprev=z;
    pause
  end
  if size(z)==size(zold)
    if z==zold
      learning=0;
    end
  end
  zold=z;
end
```

A.2 MATLAB CODE EXAMPLE FOR ARTMAP

This ARTMAP routine learns to recognize the fonts for letters 'A', 'B', and 'C' shown in Fig. 1.6 of Chapter 1. The system is first trained with these exemplars, then noisy versions of the input patterns are given, and the system classifies them. The program provides a graphic output showing the noisy test pattern and indicating whether it has been classified into *Class* 1 (letter 'A'), *Class* 2 (letter 'B'), *Class* 3, (letter 'C'), or *Class* "Don't Know".

The program is called from the MATLAB prompt using the command

```
>> artmap(rhobar,L,noise)
```

where *rhobar* corresponds to parameter $\overline{\rho_a}$ discussed in Section 1.3, and noise is a number between '0' and '1' which controls how much the noisy input patterns are degraded for the test stage. If $noise = 0$ no degradation appears and if $noise = 1$ input patterns are pure noise.

The main program consists of the routines "*artmap*" and "*test_artmap*" whose code is given next, and the auxiliary functions "*initI*", "*initb*", "*degrade*", "*msq*", and "*art1plot*" given at the end of this Appendix.

File: "artmap.m"

```
function artmap(rhobar,L,noise)
epsilon=0.0001;
no=100;
n=200;
np=18;
Mb=3;
M=1;
Io=initI;
Ic=1-Io;
I=[Io;Ic];
b=initb;
z=ones(n,1);
w=ones(Mb,1);
zold=z;
wold=w;
learning=1;
iterations=0;
while learning==1
  iterations=iterations+1;
  for pattern=1:np
    rho=rhobar;
    ok=0;
    for j=1:M
      T(j)=norm(min(I(:,pattern),z(:,j)),1)/(L-1+norm(z(:,j),1));
    end
    while ok==0
      [maxT,J]=max(T);
      if rho*norm(I(:,pattern),1) <= norm(min(I(:,pattern),z(:,J)),1)
        [maxW,K]=max(b(:,pattern));
```

```
          if w(K,J) == 1
            ok=1;
          else
         rho=norm(min(I(:,pattern),z(:,J)),1)/norm(I(:,pattern),1)+epsilon;
            T(J) = -1;
            ok=0;
          end
        else
          T(J) = -1;
        end
      end
      z(:,J)=min(I(:,pattern),z(:,J));
      w(:,J)=min(b(:,pattern),w(:,J));
      if J==M
        M=M+1;
        z=[z ones(n,1)];
        w=[w ones(Mb,1)];
      end
    end
    if size(z)==size(zold)
      if z==zold
        if size(w)==size(wold)
          if w==wold
            learning=0;
          end
        end
      end
    end
    zold=z;
    wold=w;
  end
  iterations
  categories=M
  test_artmap(Io,z,w,rhobar,L,10,10,noise);
```

File: "test_artmap.m"

```
function test_artmap(Io,z,w,rhobar,L,n1,n2,noise)

Io=degrade(Io,noise);
Ic=1-Io;
I=[Io;Ic];
rho=rhobar;
[n np]=size(I);
[Mb M]=size(w);
  for pattern=1:np
    ok=0;
    for j=1:M
      T(j)=norm(min(I(:,pattern),z(:,j)),1)/(L-1+norm(z(:,j),1));
    end
    while ok==0
      [maxT,J]=max(T);
      if rho*norm(I(:,pattern),1) <= norm(min(I(:,pattern),z(:,J)),1)
        ok=1;
      else
        T(J) = -1;
      end
    end
    [maxW,K]=max(w(:,J));
    clf
```

```
    art1plot(msq(n1,n2,Io(:,pattern))');
    if J==M
       ss=sprintf('Pattern Number:   %d, Predicted Class is:   Do not
know',pattern);
       text(0,0,ss)
    else
       ss=sprintf('Pattern Number:   %d, Predicted Class is:
%d',pattern,K);
       text(0,0,ss)
    end
    pause
  end
```

A.3 MATLAB CODE EXAMPLE FOR FUZZY-ART

This program performs the example discussed in Fig. 1.14 of Chapter 1. It takes as input a set of points $(x,y) \in \mathcal{R}^2$ inside the unit square and clusters them into categories depending on the values given for parameters ρ and α. After training is completed the program shows in a graphic window the boxes for each resulting category, and plots the input points using a different character depending on the category it has been assigned to.

Before running this Fuzzy-ART program one must run the program "*genpun*" which generates random training points and saves them in a file 'points.mat'. Program "*genpun*" is called from the MATLAB prompt using

```
>> genpun(np,nc)
```

where np is the total number of random points to be generated, and nc is the number of centers around which the random points should be generated. For example, for the training set shown in Fig. 1.13 there were $np = 100$ total points (circles) around $nc = 4$ center points (crosses). Once the training set is generated, one can run the command

```
>> fart(rho,alpha)
```

The starting routine is "*fart*" which calls the main Fuzzy-ART routine "*fuzzy*". The code for these two routines is given next, while the auxiliary functions "*genpun*", "*aug*", and "*rectangle*" are given at the end of this Appendix.

File: "fart.m"

```
function fart(rho,alpha)

aa=fuzzy(rho,alpha);
load wij;
nc=size(wij);
ncat=nc(1)-1;
load points;
np=size(points);
```

```
npunt=np(1);
sym=sprintf('.ox sdv^<>ph');
ns=size(sym);
nsym=ns(2);
col=sprintf('ymcrgbwk');
clf
h=gcf;
set(h,'name','Fuzzy ART');
hold;
for cnp=1:1:npunt
  if aa(cnp) < 13
    plot(points(cnp,1),points(cnp,2),sym(aa(cnp)));
  else
    plot(points(cnp,1),points(cnp,2),'x');
  end
end
for cc=1:1:ncat
   rectang(wij(cc,1),wij(cc,3),1-wij(cc,2),1-wij(cc,4));
end
hold;
axis('equal');
```

File: "fuzzy.m"

```
function map=fuzzy(rho,alpha)

F1=4;
CAT=0;
wij = ones(1,F1);
wijold=1;
load points;
ns=size(points);
np=ns(1);
niter=1
while 1>0,
 for point=1:1:np
    a(1)=points(point,1);
    a(2)=1-points(point,1);
    a(3)=points(point,2);
    a(4)=1-points(point,2);
    for j=1:CAT+1
      T(j) = norm(min(a,wij(j,:)),1)/(alpha+norm(wij(j,:),1));
    end
    while 1>0,
        [Tmax,Jmax]=max(T);
        if norm(min(a,wij(Jmax,:)),1) >= rho*norm(a,1)
           map(point)=Jmax;
           break;
        end
        T(Jmax)=0;
    end
    wij(Jmax,:)=min(a,wij(Jmax,:));
    if Jmax==CAT+1
      CAT=CAT+1;
      wij = aug(wij,1);
    end
 end
 if size(wij)==size(wijold)
   if wij == wijold
```

```
        break;
    end
end
wijold=wij;
ss=sprintf('niter=%d CAT=%d',niter,CAT)
niter = niter+1;
end
save wij wij -ascii
save map map -ascii
```

A.4 MATLAB CODE EXAMPLE FOR FUZZY-ARTMAP

This program executes the example discussed in Fig. 1.17 known as the problem "*Learning to Tell Two Spirals Apart*". It first generates the 194 training points of the two spirals using eqs. (1.52) and (1.53), and then trains the Fuzzy-ARTMAP system using the values for parameters $\overline{\rho_a}$ and α. Once training is complete the $\mathcal{R}^2$ unit square is partitioned into a $tnp \times tnp$ grid, and the center point of each square in this grid is classified as either belonging to the region represented by the first or the second spiral. The program provides a graphic output in which each square is assigned a color depending on the region it has been classified and draws on top the set of training points of the two spirals. The results shown in Fig. 1.17 were obtained using this program.

The program is run from the MATLAB prompt using the command

```
>> fartmap(tnp,rho,alpha)
```

Program "*fartmap*" is a starting routine which calls the main routine "*trainfam*". The code for these two routines is given next, while the code for the auxiliary functions "*spitrain*" and "*aug*" is given at the end of this Appendix.

File: "fartmap.m

```
function fartmap(tnp,rho,alpha)

figure(1)
aa=trainfam(tnp,rho,alpha);
xx=0:1/tnp:1;
yy=0:1/tnp:1;
clf
h=gcf;
axis([0 1 0 1]);
set(h,'colormap',cool);
set(h,'name','Fuzzy ARTMAP');
image(xx,yy,aa);
view(2);
d=1/(2*tnp);
axis([0-d 1+d 0-d 1+d]);
axis('equal');
axis('manual');
hold;
spiral=spitrain;
spi=size(spiral);
```

```
npi=spi(1);
for nsp=1:2:npi
 plot(spiral(nsp,1),spiral(nsp,2),'o')
 plot(spiral(nsp+1,1),spiral(nsp+1,2),'*')
end
axis off
hold;
```

File: "trainfam.m"

```
function map=trainfam(TNP,rho_init,alpha)

F1=4;
F2max=10000;
F2b=2;
CAT=0;
epsilon=0.01;
rho_ab=0.8;
wij = ones(1,F1);
wjk = ones(1,F2b);
wijold=1;
wjkold=1;
spiral=spitrain;
strain=size(spiral);
ntrain=strain(1);
niter=1
while 1>0,
   for nt=1:ntrain;
    a(1)=spiral(nt,1);
    a(2)=1-a(1);
    a(3)=spiral(nt,2);
    a(4)=1-a(3);
    b(1)=spiral(nt,3);
    b(2)=1-b(1);
    rhoa = rho_init;
    for j=1:CAT+1
      T(j) = norm(min(a,wij(j,:)),1)/(alpha+norm(wij(j,:),1));
    end
    while 1>0,
      while 1>0,
        [Tmax,Jmax]=max(T);
        if norm(min(a,wij(Jmax,:)),1) >= rhoa*norm(a,1)
          break;
        end
        T(Jmax)=0;
      end
      if norm(min(b,wjk(Jmax,:)),1) >= rho_ab
        break;
      end
      rhoa = norm(min(a,wij(Jmax,:)),1)/(F1/2)+epsilon;
      T(Jmax)=0;
    end
    wij(Jmax,:)=min(a,wij(Jmax,:));
    wjk(Jmax,:)=min(b,wjk(Jmax,:));
    if Jmax==CAT+1
      CAT=CAT+1;
      wij = aug(wij,1);
      wjk = aug(wjk,1);
```

```
          end
       rhoa=rho_init;
    end
    if size(wij)==size(wijold)
      if wij == wijold
        if size(wjk)==size(wjkold)
          if wjk == wjkold
            break;
          end
        end
      end
    end
    wijold=wij;
    wjkold=wjk;
    ss=sprintf('niter=%d CAT=%d',niter,CAT)
    niter = niter+1;
  end
  wij
  wjk
  CAT,niter
  for jy=1:TNP+1
    for ix=1:TNP+1
      a(1)=(ix-0.5)/TNP;
      a(2)=1-a(1);
      a(3)=(jy-0.5)/TNP;
      a(4)=1-a(3);
      for j=1:CAT
        T(j) = norm(min(a,wij(j,:)),1)/(alpha+norm(wij(j,:),1));
      end
      T(CAT+1)= -1;
      [Tmax,Jmax]=max(T);
      [bmax,Kmax]=max(wjk(Jmax,:));
      if Kmax==2
        map(jy,ix)=0;
      else
        map(jy,ix)=100;
      end
    end
  end
```

A.5 AUXILIARY FUNCTIONS

File: "draw_status.m"

```
function draw_status(M,Mold,pattern,np,J,Jold,I,z,zold,n1,n2,first,iter)

  clf
  subplot(2,M+2,M+2+1)
  axis('equal')
  axis off
  x='';
  for(i=1:n1)
    x=[x,I(n2*(i-1)+1:n2*i,pattern)];
  end
  art1plot(x);
  ss=sprintf('Iteration:   %d,   Pattern:   %d',iter,pattern);
  text(0,-0.5,ss)
  for(j=1:M)
```

```
      subplot(2,M+2,M+2+j+2)
      x='';
      for i=1:n1
        x=[x,z(n2*(i-1)+1:n2*i,j)];
      end
      art1plot(x);
      if j==J
        hold on
        plot([0.5,0.5],[0.5,n2+0.5],'y')
        plot([0.5,n1+0.5],[n2+0.5,n2+0.5],'y')
        plot([n1+0.5,n1+0.5],[n2+0.5,0.5],'y')
        plot([n1+0.5,0.5],[0.5,0.5],'y')
        hold off
      end
    end
    if first==1
      break;
    end
    pold=pattern-1;
    if pold==0
      pold=np;
    end
    subplot(2,M+2,1)
    axis('equal')
    axis off
    x='';
    for(i=1:n1)
      x=[x,I(n2*(i-1)+1:n2*i,pold)];
    end
    art1plot(x);
    for(j=1:Mold)
      subplot(2,M+2,j+2)
      x='';
      for i=1:n1
        x=[x,zold(n2*(i-1)+1:n2*i,j)];
      end
      art1plot(x);
      if j==Jold
        hold on
        plot([0.5,0.5],[0.5,n2+0.5],'y')
        plot([0.5,n1+0.5],[n2+0.5,n2+0.5],'y')
        plot([n1+0.5,n1+0.5],[n2+0.5,0.5],'y')
        plot([n1+0.5,0.5],[0.5,0.5],'y')
        hold off
      end
    end
end
```

File: "art1plot.m"

```
function art1plot(x)

[n,m]=size(x);
x=(1-x)*100;
image(x)
axis('off')
axis('equal')
axis([0.5 m+0.5 0.5 n+0.5])
colormap('gray')
```

File: "initI.m"

```
function I=initI
I='';
x=[0 0 0 0 0 1 0 0 0 0   0 0 0 0 1 1 1 0 0 0   0 0 0 0 1 0 1 0 0 0
   0 0 0 1 1 0 1 1 0 0   0 0 0 1 0 0 0 1 0 0   0 0 1 1 0 0 0 1 1 0
   0 0 1 0 0 0 0 0 1 0   0 1 1 1 1 1 1 1 1 1   0 1 0 0 0 0 0 0 0 1
   0 1 0 0 0 0 0 0 0 1];
I=[I,x'];
x=[0 0 0 0 0 0 0 0 1 1   0 0 0 0 0 0 0 1 0 1   0 0 0 0 0 0 1 0 0 1
   0 0 0 0 0 1 0 0 0 1   0 0 0 0 1 0 0 0 0 1   0 0 0 1 0 0 0 0 0 1
   0 0 1 0 0 0 0 0 0 1   0 1 1 1 1 1 1 1 1 1   0 1 0 0 0 0 0 0 0 1
   1 1 0 0 0 0 0 0 0 1];
I=[I,x'];
x=[0 0 0 0 0 0 0 0 0 0   0 0 0 0 0 0 0 0 0 0   0 0 0 0 0 0 0 0 0 0
   0 0 0 0 1 1 0 1 0 0   0 0 0 1 0 0 1 1 0 0   0 0 1 0 0 0 0 1 0 0
   0 0 1 0 0 0 0 1 0 0   0 0 1 0 0 0 0 1 0 0   0 0 0 1 0 0 1 1 0 0
   0 0 0 0 1 1 0 1 1 0];
I=[I,x'];
x=[0 0 0 0 0 0 0 0 0 0   0 0 0 0 0 0 0 0 0 0   0 0 0 0 1 1 0 0 0 0
   0 0 0 1 0 0 1 0 0 0   0 0 0 0 0 0 0 1 0 0   0 0 0 0 1 1 0 1 0 0
   0 0 0 1 0 0 1 1 0 0   0 0 1 0 0 0 0 1 0 0   0 0 0 1 0 0 1 1 0 0
   0 0 0 0 1 1 0 1 1 0];
I=[I,x'];
x=[0 0 0 0 0 0 0 0 0 0   0 0 0 0 0 0 0 0 0 0   0 0 0 0 0 0 0 0 0 0
   0 0 0 1 1 0 0 0 0 1   0 0 1 0 0 1 0 0 1 0   0 1 0 0 0 0 1 1 0 0
   0 1 0 0 0 0 1 0 0 0   0 1 0 0 0 0 1 1 0 0   0 0 1 0 0 1 0 0 1 0
   0 0 0 1 1 0 0 0 0 1];
I=[I,x'];
x=[0 0 0 0 0 0 0 0 0 1   0 0 0 0 0 0 0 0 1 1   0 0 0 0 0 0 0 1 1 1
   0 0 0 0 0 0 1 1 1 1   0 0 0 0 0 1 1 1 1 1   0 0 0 0 1 1 1 1 1 1
   0 0 0 1 1 1 1 1 1 1   0 0 1 1 1 1 1 1 1 1   0 1 1 1 1 1 1 1 1 1
   1 1 1 1 1 1 1 1 1 1];
I=[I,x'];
x=[0 1 1 1 1 1 1 0 0   0 0 1 0 0 0 0 0 1 0   0 0 1 0 0 0 0 0 1 0
   0 0 1 0 0 0 0 0 1 0   0 0 1 1 1 1 1 1 0 0   0 0 1 0 0 0 0 0 1 0
   0 0 1 0 0 0 0 0 0 1   0 0 1 0 0 0 0 0 0 1   0 0 1 0 0 0 0 0 1 0
   0 1 1 1 1 1 1 1 0 0];
I=[I,x'];
x=[0 0 0 1 1 1 1 1 0 0   0 0 0 0 1 0 0 0 1 0   0 0 0 0 1 0 0 0 1 0
   0 0 0 1 0 0 0 0 1 0   0 0 0 1 1 1 1 1 0 0   0 0 1 0 0 0 0 0 1 0
   0 0 1 0 0 0 0 0 0 1   0 1 0 0 0 0 0 0 0 1   0 1 0 0 0 0 0 0 1 0
   1 1 1 1 1 1 1 1 0 0];
I=[I,x'];
x=[0 0 0 0 0 0 0 0 0 0   0 0 1 0 0 0 0 0 0 0   0 0 1 0 0 0 0 0 0 0
   0 0 1 0 0 0 0 0 0 0   0 0 1 0 1 1 0 0 0 0   0 0 1 1 0 0 1 0 0 0
   0 0 1 0 0 0 0 1 0 0   0 0 1 0 0 0 0 1 0 0   0 0 1 1 0 0 1 0 0 0
   0 0 1 0 1 1 0 0 0 0];
I=[I,x'];
x=[0 0 0 0 0 0 0 0 0 0   0 0 0 1 0 0 0 0 0 0   0 0 0 1 0 0 0 0 0 0
   0 0 0 1 0 0 0 0 0 0   0 0 1 1 1 1 0 0 0 0   0 0 1 0 0 0 1 0 0 0
   0 0 1 0 0 0 0 1 0 0   0 1 1 0 0 0 0 1 0 0   0 1 0 1 0 0 1 0 0 0
   0 1 0 0 1 1 0 0 0 0];
I=[I,x'];
x=[0 0 0 1 1 0 0 0 0 0   0 0 1 0 0 1 0 0 0 0   0 1 0 0 0 0 1 0 0 0
   0 1 0 0 0 0 1 0 0 0   0 1 0 1 1 1 0 0 0 0   0 1 0 0 0 0 1 0 0 0
   0 1 0 0 0 0 1 0 0 0   0 1 1 0 0 1 0 0 0 0   0 1 0 1 1 0 0 0 0 0
   0 1 0 0 0 0 0 0 0 0];
I=[I,x'];
x=[1 1 1 1 0 0 0 0 0 0   1 1 1 1 1 0 0 0 0 0   1 1 1 1 1 0 0 0 0 0
```

```
   1 1 1 1 0 0 0 0 0   1 1 1 1 1 1 0 0 0 0   1 1 1 1 1 1 1 1 1 0
   1 1 1 1 1 1 1 1 1 1   1 1 1 1 1 1 1 1 1 0   1 1 1 1 1 1 0 0 0 0
   1 1 1 0 0 0 0 0 0 0];
I=[I,x'];
x=[0 0 0 0 1 1 1 1 0 1   0 0 0 1 0 0 0 0 1 1   0 0 1 0 0 0 0 0 0 1
   0 1 0 0 0 0 0 0 0 0   0 1 0 0 0 0 0 0 0 0   0 1 0 0 0 0 0 0 0 0
   0 1 0 0 0 0 0 0 0 0   0 0 1 0 0 0 0 0 0 1   0 0 0 1 0 0 0 0 1 1
   0 0 0 0 1 1 1 1 0 1];
I=[I,x'];
x=[0 0 0 0 1 1 1 0 0 1   0 0 0 1 0 0 0 1 1 1   0 0 0 1 0 0 0 0 0 1
   0 0 1 0 0 0 0 0 0 0   0 0 1 0 0 0 0 0 0 0   0 1 0 0 0 0 0 0 0 0
   0 1 0 0 0 0 0 0 0 0   0 1 0 0 0 0 0 0 0 1   0 1 1 0 0 0 0 1 1 1
   0 0 0 1 1 1 1 0 0 1];
I=[I,x'];
x=[0 0 0 0 0 0 0 0 0 0   0 0 0 0 0 0 0 0 0 0   0 0 0 0 0 0 0 0 0 0
   0 0 0 0 0 0 0 0 0 0   0 0 0 0 0 1 1 0 0 0   0 0 0 0 1 0 0 1 0 0
   0 0 1 0 0 0 0 0 0 0   0 0 0 1 0 0 0 0 0 0   0 0 0 0 1 0 0 1 0 0
   0 0 0 0 0 1 1 0 0 0];
I=[I,x'];
x=[0 0 0 0 0 0 0 0 0 0   0 0 0 0 0 0 0 0 0 0   0 0 0 0 0 0 0 0 0 0
   0 0 0 0 0 0 0 0 0 0   0 0 0 0 1 1 1 0 0 0   0 0 0 1 0 0 0 1 0 0
   0 0 1 0 0 0 0 0 0 0   0 0 0 1 0 0 0 0 0 0   0 0 0 1 0 0 0 1 0 0
   0 0 0 0 1 1 1 0 0 0];
I=[I,x'];
x=[0 0 0 1 0 0 1 0 0 0   0 0 0 0 1 1 0 0 0 0   0 0 0 0 1 1 0 0 0 0
   0 0 0 1 0 0 1 0 0 0   0 0 0 1 0 0 1 0 0 0   0 0 1 0 0 0 0 1 0 0
   0 0 1 0 0 0 0 1 0 0   0 0 0 1 0 0 1 0 0 0   0 0 0 1 0 0 1 0 0 0
   0 0 0 0 1 1 0 0 0 0];
I=[I,x'];
x=[0 0 0 0 0 0 1 1 1   0 0 0 0 0 1 1 1 1 1   0 0 0 0 1 1 1 1 1 1
   0 0 0 1 1 1 1 1 1 1   0 0 1 1 1 1 1 1 1 1   0 0 1 1 1 1 1 1 1 1
   0 0 0 1 1 1 1 1 1 1   0 0 0 0 1 1 1 1 1 1   0 0 0 0 0 1 1 1 1 1
   0 0 0 0 0 0 0 1 1 1];
I=[I,x'];
```

File: "initb.m"

```
function b=initb

b=[1 0 0;1 0 0;1 0 0;1 0 0;1 0 0;1 0 0;
   0 1 0;0 1 0;0 1 0;0 1 0;0 1 0;0 1 0;
   0 0 1;0 0 1;0 0 1;0 0 1;0 0 1;0 0 1];
b=b';
```

File: "degrade.m"

```
function a=degrade(I,noise)

if noise<0
  Error('Noise must be between 0 and 1')
end
if noise>1
  Error('Noise must be between 0 and 1')
end
[n np]=size(I);
for i=1:n
```

```
    for j=1:np
      if rand(1,1)<noise
         a(i,j)=round(rand(1,1));
      else
         a(i,j)=I(i,j);
      end
    end
end
```

File: "msq.m"

```
function x=msq(n1,n2,I)

  if size(I) = n1*n2
    Error('Error in "msq.m"');
  end
  x='';
  for i=1:n1
    x=[x,I((i-1)*n2+1:i*n2,1)];
  end
```

File: "genpun.m"

```
function genpun(np,nc)

clusx=rand(nc,1);
clusy=rand(nc,1);
figure(1)
for point=1:1:np
    ac=rand(1);
    dcx=0.1*randn(1);
    dcy=0.1*randn(1);
    for cc=1:1:nc
        if ac< cc/nc
            if ac>(cc-1)/nc
                punt(point,1)=clusx(cc,1)+dcx;
                punt(point,2)=clusy(cc,1)+dcy;
            end
        end
    end
end
ma=max(max(punt));
mi=min(min(punt));
punt=(punt-mi)/(ma-mi);
clusx=(clusx-mi)/(ma-mi);
clusy=(clusy-mi)/(ma-mi);
plot(clusx,clusy,'+')
save points punt -ascii
hold
plot(punt(:,1),punt(:,2),'o')
hold
```

File: "aug.m"

```
function aa=aug(a,n)

[lines cols]=size(a);
u=ones(1,cols);
for i=1:n
  a = [a;u];
end
aa = a;
```

File: "rectang.m"

```
function rectang(x1,y1,x2,y2)

line([x1,x2],[y1,y1]);
line([x2,x2],[y1,y2]);
line([x2,x1],[y2,y2]);
line([x1,x1],[y2,y1]);
```

File: "spitrain.m"

```
function spiral=spitrain

for n=1:97
    alpha(n)=pi*(n-1)/16;
    r(n)=0.4*((105-n)/104);
    a(2*n-1,1)=r(n)*sin(alpha(n))+0.5;
    a(2*n-1,2)=r(n)*cos(alpha(n))+0.5;
    b(2*n-1,1)=1;
    a(2*n,1)=1-a(2*n-1,1);
    a(2*n,2)=1-a(2*n-1,2);
    b(2*n,1)=0;
    spiral=[a,b];
end
```

Appendix B
Computational Equivalence of the Original ART1 and the Modified ART1m Models

In the original ART1 paper [Carpenter, 1987] the architecture is mathematically described as sets of Short Term Memory (STM) and Long Term Memory (LTM) time domain nonlinear differential equations. The STM differential equations describe the evolution and interactions between processing units or neurons of the system, while the LTM differential equations describe how the interconnection weights change in time as a function of the state of the system. The time constants associated to the LTM differential equations are much slower than those associated to the STM differential equations. A valid assumption is to make the STM differential equations settle instantaneously to their corresponding steady state, and consider only the dynamics of the LTM differential equations. In this case, the STM differential equations must be substituted by nonlinear algebraic equations that describe the corresponding steady state of the system. Furthermore, Carpenter and Grossberg also introduced the *Fast Learning* mode of the ART1 architecture, in which the LTM differential equations are also substituted by their corresponding steady-state nonlinear algebraic equations. Thus, the ART1 architecture originally modeled as a dynamically evolving collection of neurons and synapses governed by time-domain differential equations, can be behaviorally modeled as the sequential application of nonlinear algebraic equations: an input pattern is given, the corresponding STM steady state is computed through the STM algebraic equations, and the system weights are updated using the corresponding LTM algebraic equations. At this point three different levels of ART1 implementations (in both software or hardware) can be distinguished:

- **Type-1, Full Model Implementation:** Both STM and LTM time-domain differential equations are realized. This implementation is the most expensive, and requires a large amount of computational power.

- **Type-2, STM Steady-State Implementation:** Only the LTM time-domain differential equations are implemented. The STM behavior is governed by nonlinear algebraic equations. This implementation requires less resources than the previous one. However, proper sequencing of STM events must be introduced artificially, which is architecturally implicit in the Type-1 implementation.

- **Type-3, Fast Learning Implementation:** This implementation is computationally the least expensive. In this case, STM and LTM events must be artificially sequenced.

Throughout the original ART1 paper [Carpenter, 1987], Carpenter and Grossberg provide rigorous demonstrations of the computational properties of the ART1 architecture. Some of these properties are concerned with *Type*-1 and *Type*-2 operations of the architecture, but most refer to the *Type*-3 model operation. From a functional point of view, i.e., when looking at the ART1 system as a black box regardless of the details of its internal operations, the system level computational properties of ART1 are fully contained in its *Fast-Learning* or *Type*-3 model. The theorems and demonstrations given by Carpenter and Grossberg [Carpenter, 1987] relating to *Type*-1 and *Type*-2 models of the system only ensure proper *Type*-3 behavior. The purpose of this Appendix is to demonstrate that the modified *Type*-3 model (ART1m) developed in Chapter 2 preserves all the *Type*-3 computational properties of the original ART1 architecture. The only functional difference between ART1 and ART1m, is the way the terms T_j are computed before competing in the Winner-Takes-All block. Therefore, the original properties and demonstrations that are not affected by the terms T_j will be automatically preserved. Such properties are, for example, the *Self-Scaling* property and the *Variable Coarseness* property tuned by the *Vigilance Parameter*. But there are other properties which are directly affected by the way the terms T_j are computed. In the remainder of this Appendix we will show that these properties remain in the ART1m architecture.

Let us define a few concepts before demonstrating that the original computational properties are preserved.

1. *Direct Access*: an input pattern **I** is said to have *Direct Access* to a learned category j if this category is the first one selected by the Winner-Takes-All F2 layer and is accepted by the *vigilance subsystem*, so that no reset occurs.

2. *Subset Template*: an input pattern **I** is said to be a *Subset Template* of a learned category $\mathbf{z}_j \equiv (z_{1j}, z_{2j}, \ldots, z_{Nj})$ if $\mathbf{I} \subset \mathbf{z}_j$. Formally,

$$z_{ij} = 0 \Rightarrow I_i = 0 \quad \forall i = 1, \ldots, N$$
$$I_i = 1 \Rightarrow z_{ij} = 1 \quad \forall i = 1, \ldots, N \quad \text{(B.1)}$$
$$\text{there might be values of } i \text{ such that} \quad I_i = 0 \text{ and} \quad z_{ij} = 1$$

APPENDIX B: COMPUTATIONAL EQUIVALENCE OF ART1 AND ART1M

3. *Superset Template*: an input pattern **I** is said to be a *Superset Template* of a learned category j if $\mathbf{z}_j \subset \mathbf{I}$.

4. *Mixed Template*: $\mathbf{z}_j$ and **I** are said to be mixed templates if neither $\mathbf{I} \subset \mathbf{z}_j$ nor $\mathbf{z}_j \subset \mathbf{I}$ are satisfied, and $\mathbf{I} \neq \mathbf{z}_j$.

5. *Uncommitted node*: an $F2$ node j is said to be uncommitted if all its weights $z_{ij}(i = 1, \ldots, N)$ preserve their initial value ($z_{ij} = 1$), i.e., node j has not yet been selected to represent any learned category.

B.1 DIRECT ACCESS TO SUBSET AND SUPERSET PATTERNS

Suppose that a learning process has produced a set of categories in the $F2$ layer. Each category j is characterized by the set of weights that connect node j in the $F2$ layer to all nodes in the $F1$ layer, i.e., $\mathbf{z}_j \equiv (z_{1j}, z_{2j}, \ldots, z_{Nj})$. Suppose that two of these categories, j_1 and j_2, are such that $\mathbf{z}_{j_1} \subset \mathbf{z}_{j_2}$ ($\mathbf{z}_{j_1}$ is a subset template of $\mathbf{z}_{j_2}$). Now consider two input patterns $\mathbf{I}^{(1)}$ and $\mathbf{I}^{(2)}$ such that,

$$\begin{aligned} \mathbf{I}^{(1)} &= \mathbf{z}_{j_1} \equiv (z_{1j_1}, z_{2j_1}, \ldots, z_{Nj_1}), \\ \mathbf{I}^{(2)} &= \mathbf{z}_{j_2} \equiv (z_{1j_2}, z_{2j_2}, \ldots, z_{Nj_2}). \end{aligned} \quad (B.2)$$

The *Direct Access to Subset and Superset* property assures that input $\mathbf{I}^{(1)}$ will have *Direct Access* to category j_1 and that input $\mathbf{I}^{(2)}$ will have *Direct Access* to category j_2. The proofs for this are as follows.

Original ART1:

Let us compute the values of T_{j_1} and T_{j_2} when the input patterns $\mathbf{I}^{(1)}$ and $\mathbf{I}^{(2)}$ are presented at the input of the system. For pattern $\mathbf{I}^{(1)}$ we will have,

$$\begin{aligned} T_{j_1} &= \frac{L|\mathbf{I}^{(1)} \cap \mathbf{z}_{j_1}|}{L - 1 + |\mathbf{z}_{j_1}|} = \frac{L|\mathbf{I}^{(1)}|}{L - 1 + |\mathbf{I}^{(1)}|} \\ T_{j_2} &= \frac{L|\mathbf{I}^{(1)} \cap \mathbf{z}_{j_2}|}{L - 1 + |\mathbf{z}_{j_2}|} = \frac{L|\mathbf{I}^{(1)}|}{L - 1 + |\mathbf{I}^{(2)}|} \end{aligned} \quad (B.3)$$

Since $|\mathbf{I}^{(1)}| < |\mathbf{I}^{(2)}|$, it is obvious that $T_{j_1} > T_{j_2}$ (remember that $L > 1$) and therefore category j_1 will become the active one. On the other hand, if input pattern $\mathbf{I}^{(2)}$ is presented at the input,

$$T_{j_1} = \frac{L|\mathbf{I}^{(2)} \cap \mathbf{z}_{j_1}|}{L-1+|\mathbf{z}_{j_1}|} = \frac{L|\mathbf{I}^{(1)}|}{L-1+|\mathbf{I}^{(1)}|}$$
$$T_{j_2} = \frac{L|\mathbf{I}^{(2)} \cap \mathbf{z}_{j_2}|}{L-1+|\mathbf{z}_{j_2}|} = \frac{L|\mathbf{I}^{(2)}|}{L-1+|\mathbf{I}^{(2)}|}$$
(B.4)

Since the function $Lx/(L-1+x)$ is an increasing function of x, it results that now $T_{j_2} > T_{j_1}$ and category j_2 will be chosen as the winner.

Modified ART1:

If pattern $\mathbf{I}^{(1)}$ is given as the input pattern we will have

$$T_{j_1} = L_A|\mathbf{I}^{(1)} \cap \mathbf{z}_{j_1}| - L_B|\mathbf{z}_{j_1}| + L_M = L_A|\mathbf{I}^{(1)}| - L_B|\mathbf{I}^{(1)}| + L_M$$
$$T_{j_2} = L_A|\mathbf{I}^{(1)} \cap \mathbf{z}_{j_2}| - L_B|\mathbf{z}_{j_2}| + L_M = L_A|\mathbf{I}^{(1)}| - L_B|\mathbf{I}^{(2)}| + L_M$$
(B.5)

Since $|\mathbf{I}^{(1)}| < |\mathbf{I}^{(2)}|$, it follows that (remember that $L_B > 0$) $T_{j_1} > T_{j_2}$. In the case pattern $\mathbf{I}^{(2)}$ is presented at the input of the network it would be,

$$T_{j_1} = L_A|\mathbf{I}^{(2)} \cap \mathbf{z}_{j_1}| - L_B|\mathbf{z}_{j_1}| + L_M = L_A|\mathbf{I}^{(1)}| - L_B|\mathbf{I}^{(1)}| + L_M$$
$$T_{j_2} = L_A|\mathbf{I}^{(2)} \cap \mathbf{z}_{j_2}| - L_B|\mathbf{z}_{j_2}| + L_M = L_A|\mathbf{I}^{(2)}| - L_B|\mathbf{I}^{(2)}| + L_M$$
(B.6)

In order to guarantee that $T_{j_2} > T_{j_1}$ the condition

$$L_A > L_B \tag{B.7}$$

has to be assured.

B.2 DIRECT ACCESS BY PERFECTLY LEARNED PATTERNS (THEOREM 1 OF ORIGINAL ART1)

This theorem, adapted to a *Type*-3 implementation, states the following

> An input pattern $\mathbf{I}$ has direct access to a node J which has perfectly learned the input pattern $\mathbf{I}$.

The proofs are as follows.

APPENDIX B: COMPUTATIONAL EQUIVALENCE OF ART1 AND ART1M

Original ART1:

In order to prove that **I** has direct access to J, we need to demonstrate that the following properties hold: (i) J is the first node to be chosen, (ii) J is accepted by the vigilance subsystem and (iii) J remains active as learning takes place. To prove property (i) we have to show that, at the start of each trial $T_J > T_j \; \forall j \neq J$. Since $\mathbf{I} = \mathbf{z}_J$,

$$T_J = \frac{L|\mathbf{I}|}{L - 1 + |\mathbf{I}|} \tag{B.8}$$

and

$$T_j = \frac{L|\mathbf{I} \cap \mathbf{z}_j|}{L - 1 + |\mathbf{z}_j|} \tag{B.9}$$

Since $\frac{Lw}{L-1+w}$ is an increasing function of w (because $L > 1$) and $|\mathbf{I}|, |\mathbf{z}_j| > |\mathbf{I} \cap \mathbf{z}_j|$, we can state,

$$T_J = \frac{L|\mathbf{I}|}{L - 1 + |\mathbf{I}|} > \frac{L|\mathbf{I} \cap \mathbf{z}_j|}{L - 1 + |\mathbf{I} \cap \mathbf{z}_j|} > \frac{L|\mathbf{I} \cap \mathbf{z}_j|}{L - 1 + |\mathbf{z}_j|} = T_j. \tag{B.10}$$

So, property (i) is always fulfilled.

Property (ii) is directly verified since $|\mathbf{I} \cap \mathbf{z}_j| = |\mathbf{I}| \geq \rho |\mathbf{I}| \; \forall \rho \in [0,1]$. Property (iii) is always verified because after node J is selected as the winning category, its weight template $\mathbf{z}_J$ will remain unchanged (because $\mathbf{z}_J|_{new} = \mathbf{I} \cap \mathbf{z}_J|_{old} = \mathbf{I} = \mathbf{z}_J|_{old}$), and consequently the inputs to the $F2$ layer T_j will remain unchanged.

Modified ART1:

In order to demonstrate that **I** has direct access to J, we have only to prove that property (i) is verified for the modified algorithm, as the proof of properties (ii) and (iii) is identical to the case of the original algorithm. To prove property (i), we have to demonstrate that

$$T_J = L_A|\mathbf{I}| - L_B|\mathbf{I}| + L_M > L_A|\mathbf{I} \cap \mathbf{z}_j| - L_B|\mathbf{z}_j| + L_M = T_j \tag{B.11}$$

Since $L_A w - L_B w + L_M$ is an increasing function of w ($L_A > L_B$), and $|\mathbf{I}|, |\mathbf{z}_j| > |\mathbf{I} \cap \mathbf{z}_j|$,

$$T_J = L_A|\mathbf{I}| - L_B|\mathbf{I}| + L_M > L_A|\mathbf{I} \cap \mathbf{z}_j| - L_B|\mathbf{I} \cap \mathbf{z}_j| + L_M > \quad (B.12)$$
$$> L_A|\mathbf{I} \cap \mathbf{z}_j| - L_B|\mathbf{z}_j| + L_M = T_j$$

B.3 STABLE CHOICES IN STM (THEOREM 2 OF ORIGINAL ART1)

Whenever an input pattern **I** is presented for the first time to the ART1 system, a set of T_j values is formed that compete in the Winner-Takes-All $F2$ layer. The winner may be reset by the *vigilance subsystem*, and a new winner appears that may also be reset, and so on until a final winner is accepted. During this search process, the T_j values that led to earlier winners are set to zero. Let us call O_j the values of T_j at the beginning of the search process, i.e., before any of them is set to zero by the vigilance subsystem. Theorem 2 of the original ART1 architecture states:

> Suppose that an $F2$ node J is chosen for STM storage instead of another node j because $O_J > O_j$. Then read-out of the top-down template preserves the inequality $T_J > T_j$ and thus confirms the choice of J by the bottom-up filter.

This theorem has only sense for a *Type*-1 implementation, because there, as a node in the $F2$ layer activates, the initial values of T_j (immediately after presenting an input pattern **I**) may be altered through the top-down *'feed-back'* connections. In a *Type*-3 description (see Fig. 2.1) the initial terms T_j remain unchanged, independently of what happens in the $F2$ layer. Therefore, this theorem is implicitly satisfied.

B.4 INITIAL FILTER VALUES DETERMINE SEARCH ORDER (THEOREM 3 OF ORIGINAL ART1)

Theorem 3 of the original ART1 architecture states that (page 92 of [Carpenter, 1987]):

> The Order Function ($O_{j_1} > O_{j_2} > O_{j_3} > \dots$) determines the order of search no matter how many times $F2$ is reset during a trial.

The proof is the same for the original ART1 and the modified ART1 (both *Type*-3) implementation[1]. If T_{j_1} is reset by the *vigilance subsystem*, the values of $T_{j_2}, T_{j_3}, \dots$ will not change. Therefore, the new order sequence is $O_{j_2} > O_{j_3} > \dots$ and the original second largest value O_{j_2} will be selected as the winner. If T_{j_2} is now set to zero, O_{j_3} is the next winner, and so on.

---
[1] However, note that the resulting ordering $\{j_1, j_2, j_3, \dots\}$ can be different for the original and for the modified architecture

APPENDIX B: COMPUTATIONAL EQUIVALENCE OF ART1 AND ART1M   197

This Theorem, although trivial in a *Type*-3 implementation, has more importance in a *Type*-1 description where the process of selecting and shutting down a winner has the consequence of altering T_j values.

B.5 LEARNING ON A SINGLE TRIAL (THEOREM 4 OF ORIGINAL ART1)

This theorem (page 93 of [Carpenter, 1987]) states the following, assuming a *Type*-3 implementation is being considered[2]:

> Suppose that an *F2* winning node J is accepted by the vigilance subsystem. Then the LTM traces z_{ij} change in such a way that T_J increases and all other T_j remain constant, thereby confirming the choice of J. In addition, the set $\mathbf{I} \cap \mathbf{z}_J$ remains constant during learning, so that learning does not trigger reset of J by the vigilance subsystem.

The proofs are as follows.

Original ART1:

According to eq. (2.3), if J is the winning category accepted by the vigilance subsystem, we have that

$$T_J = \frac{L|\mathbf{I} \cap \mathbf{z}_J|}{L - 1 + |\mathbf{z}_J|} \qquad (B.13)$$

This is the T_J value before learning takes place. After updating the weights (see Fig. 2.1(b)),

$$\mathbf{z}_J(new) = \mathbf{I} \cap \mathbf{z}_J(old) \qquad (B.14)$$

and the new T_J value is given by,

$$\begin{aligned} T_J(new) &= \frac{L|\mathbf{I} \cap \mathbf{z}_J(new)|}{L - 1 + |\mathbf{z}_J(new)|} = \frac{L|\mathbf{I} \cap \mathbf{I} \cap \mathbf{z}_J(old)|}{L - 1 + |\mathbf{I} \cap \mathbf{z}_J(old)|} \geq \\ &\geq \frac{L|\mathbf{I} \cap \mathbf{z}_J(old)|}{L - 1 + |\mathbf{z}_J(old)|} = T_J(old) \end{aligned} \qquad (B.15)$$

---

[2]In the original ART1 paper[Carpenter, 1987] a more sophisticated demonstration for this theorem is provided. The reason is that there the demonstration is performed for a *Type*-1 description of ART1.

Note that by eq. (B.14),

$$\mathbf{I} \cap \mathbf{z}_J(new) = \mathbf{I} \cap \mathbf{I} \cap \mathbf{z}_J(old) = \mathbf{I} \cap \mathbf{z}_J(old) \qquad (B.16)$$

Since the only weights that are updated are those connected to the winning (and accepted) node J, all other $T_j|_{j \neq J}$ values remain unchanged. Therefore, it can be concluded, by eq. (B.15), that learning confirms the choice of J and that, by eq. (B.16), the set $\mathbf{I} \cap \mathbf{z}_J$ remains constant.

Modified ART1:

In this case, if J is the winning category accepted by the *vigilance subsystem*, by eq. (2.4) we have that

$$T_J = L_A |\mathbf{I} \cap \mathbf{z}_J| - L_B |\mathbf{z}_J| + L_M \qquad (B.17)$$

The update rule is the same as before (see Fig. 2.1(b)), therefore

$$\mathbf{z}_J(new) = \mathbf{I} \cap \mathbf{z}_J(old) \qquad (B.18)$$

and the new T_J value is given now by,

$$\begin{aligned} T_J(new) &= L_A |\mathbf{I} \cap \mathbf{z}_J(old)| - L_B |\mathbf{I} \cap \mathbf{z}_J(old)| + L_M \geq \\ &\geq L_A |\mathbf{I} \cap \mathbf{z}_J(old)| - L_B |\mathbf{z}_J(old)| + L_M = T_J(old) \end{aligned} \qquad (B.19)$$

Like before, learning confirms the choice of J, and by eq. (B.18) the set $\mathbf{I} \cap \mathbf{z}_J$ remains constant as well.

B.6 STABLE CATEGORY LEARNING (THEOREM 5 OF ORIGINAL ART1)

Suppose an arbitrary list (finite or infinite) of binary input patterns is presented to an ART1 system. Each template set $\mathbf{z}_j \equiv (z_{1j}, z_{2j}, \ldots, z_{Nj})$ is updated every time category j is selected by the Winner-Takes-All $F2$ layer and accepted by the vigilance subsystem. Some of these times template $\mathbf{z}_j$ might be changed, and some others it might stay unchanged. Let us call the times $\mathbf{z}_j$ suffers a change $t_1^{(j)} < t_2^{(j)} < \cdots < t_{r_j}^{(j)}$. Since vector (or template) $\mathbf{z}_j$ has N components (initially set to '1'), and by eq. (B.14), each component can only change from

'1' to '0' but not from '0' to '1', it follows that template z_j can, at the most, suffer $N-1$ changes[3],

$$r_j \leq N \tag{B.20}$$

Since template z_j will remain unchanged after time $t_{r_j}^{(j)}$, it is concluded that the complete LTM memory will suffer no change after time

$$t_{learn} = \max_j \{t_{r_j}^{(j)}\} \tag{B.21}$$

If there is a finite number of nodes in the $F2$ layer t_{learn} has a finite value, and thus learning completes after a finite number of time steps.

All this is true for both, the original and the modified ART1 architecture, and therefore the following theorem (page 95 of [Carpenter, 1987]) is valid for the two algorithms:

> In response to an arbitrary list of binary input patterns, all LTM traces $z_{ij}(t)$ approach limits after a finite number of learning trials. Each template set z_j remains constant except for at most N times $t_1^{(j)} < t_2^{(j)} < \cdots < t_{r_j}^{(j)}$ at which it progressively loses elements, leading to the
>
> Subset Recoding Property: $z_j(t_1^{(j)}) \supset z_j(t_2^{(j)}) \supset \cdots \supset z_j(t_{r_j}^{(j)})$. (B.22)
>
> The LTM traces $z_{ij}(t)$ such that $i \notin z_j(t_{r_j}^{(j)})$ decrease to zero. The LTM traces $z_{ij}(t)$ such that $i \in z_j(t_{r_j}^{(j)})$ remain always at '1'. The LTM traces such that $i \in z_j(t_k^{(j)})$ but $i \notin z_j(t_{k+1}^{(j)})$ stay at '1' for times $t \leq t_k^{(j)}$ but will change to and stay at '0' for times $t \geq t_{k+1}^{(j)}$.

B.7 DIRECT ACCESS AFTER LEARNING SELF-STABILIZES (THEOREM 6 OF ORIGINAL ART1)

Assuming $F2$ has a finite number of nodes, the present theorem (page 98 of [Carpenter, 1987]) states the following:

> After recognition learning has self-stabilized in response to an arbitrary list of binary input patterns, each input pattern **I** either has direct access to the node j

---

[3] As mentioned at the end of Chapter 2, the empty template is not valid for the original ART1 system. Therefore, eq. (B.20) can be changed to $r_j \leq N-1$ for ART1, but not for ART1m.

which possesses the largest subset template with respect to **I**, or **I** cannot be coded by any node of *F2*. In the latter case, *F2* contains no uncommitted nodes.

Since learning has already stabilized **I** can be coded only by a node j whose template $\mathbf{z}_j$ is a subset template with respect to **I**. Otherwise, after j becomes active, the set $\mathbf{z}_j$ would contract to $\mathbf{z}_j \cap \mathbf{I}$, thereby contradicting the hypothesis that learning has already stabilized. Thus if **I** activates any node other than one with a subset template, that node must be reset by the *vigilance subsystem*. For the remainder of the proof, let J be the first *F2* node activated by **I**. We need to show that if $\mathbf{z}_J$ is a subset template, then it is the subset template with the largest O_J; and if it is not a subset template, then all subset templates activated on that trial will be reset by the vigilance subsystem. To proof these two steps we need to differentiate between the original ART1 and the modified one.

Original ART1:

If J and j are nodes with subset templates with respect to **I**, then

$$O_j = \frac{L|\mathbf{z}_j|}{L-1+|\mathbf{z}_j|} < O_J = \frac{L|\mathbf{z}_J|}{L-1+|\mathbf{z}_J|} \tag{B.23}$$

Since $\frac{L|\mathbf{z}_j|}{L-1+|\mathbf{z}_j|}$ is an increasing function of $|\mathbf{z}_j|$,

$$|\mathbf{z}_j| < |\mathbf{z}_J| \tag{B.24}$$

and,

$$R_j = \frac{|\mathbf{I} \cap \mathbf{z}_j|}{|\mathbf{I}|} = \frac{|\mathbf{z}_j|}{|\mathbf{I}|} < R_J = \frac{|\mathbf{I} \cap \mathbf{z}_J|}{|\mathbf{I}|} = \frac{|\mathbf{z}_J|}{|\mathbf{I}|} \tag{B.25}$$

Once activated, a node k will be reset if $R_k < \rho$. Therefore, if J is reset ($R_J < \rho$), then all other nodes with subset templates will be reset as well ($R_j < \rho$).

Now suppose that J, the first activated node, does not have a subset template with respect to **I** ($|\mathbf{I} \cap \mathbf{z}_J| < |\mathbf{z}_J|$), but that another node j with a subset template is activated in the course of search. We need to show that $|\mathbf{I} \cap \mathbf{z}_j| = |\mathbf{z}_j| < \rho|\mathbf{I}|$, so that j is reset. We know that,

$$O_j = \frac{L|\mathbf{z}_j|}{L-1+|\mathbf{z}_j|} < O_J = \frac{L|\mathbf{I} \cap \mathbf{z}_J|}{L-1+|\mathbf{z}_J|} < \frac{L|\mathbf{z}_J|}{L-1+|\mathbf{z}_J|} \tag{B.26}$$

APPENDIX B: COMPUTATIONAL EQUIVALENCE OF ART1 AND ART1M

which implies that $|\mathbf{z}_j| < |\mathbf{z}_J|$. Since J cannot be chosen, it has to be reset by the *vigilance subsystem*, which means that $|\mathbf{I} \cap \mathbf{z}_J| < \rho|\mathbf{I}|$. Therefore,

$$\frac{|\mathbf{z}_j|}{L-1+|\mathbf{z}_j|} < \frac{|\mathbf{I} \cap \mathbf{z}_J|}{L-1+|\mathbf{z}_J|} < \frac{\rho|\mathbf{I}|}{L-1+|\mathbf{z}_J|} < \frac{\rho|\mathbf{I}|}{L-1+|\mathbf{z}_j|} \qquad (B.27)$$

which implies that,

$$|\mathbf{I} \cap \mathbf{z}_j| = |\mathbf{z}_j| < \rho|\mathbf{I}| \qquad (B.28)$$

Modified ART1:

If J and j are nodes with subset templates with respect to $\mathbf{I}$, then

$$O_j = L_A|\mathbf{z}_j| - L_B|\mathbf{z}_j| + L_M < O_J = L_A|\mathbf{z}_J| - L_B|\mathbf{z}_J| + L_M \qquad (B.29)$$

Since $(L_A - L_B)|\mathbf{z}_j|$ is an increasing function of $|\mathbf{z}_j|$,

$$|\mathbf{z}_j| < |\mathbf{z}_J| \qquad (B.30)$$

and,

$$R_j = \frac{|\mathbf{I} \cap \mathbf{z}_j|}{|\mathbf{I}|} = \frac{|\mathbf{z}_j|}{|\mathbf{I}|} < R_j = \frac{|\mathbf{I} \cap \mathbf{z}_J|}{|\mathbf{I}|} = \frac{|\mathbf{z}_J|}{|\mathbf{I}|} \qquad (B.31)$$

Therefore, if J is reset ($R_J < \rho$), then all other nodes with subset templates will be reset as well ($R_j < \rho$).

Now suppose that J, the first activated node, does not have a subset template with respect to $\mathbf{I}$ ($|\mathbf{I} \cap \mathbf{z}_J| < |\mathbf{z}_J|$), but that another node j with a subset template is activated in the course of search. We need to show that $|\mathbf{I} \cap \mathbf{z}_J| = |\mathbf{z}_j| < \rho|\mathbf{I}|$, so that j is reset. We know that,

$$\begin{aligned} O_j &= (L_A - L_B)|\mathbf{z}_j| + L_M < O_J = L_A|\mathbf{I} \cap \mathbf{z}_J| - L_B|\mathbf{z}_J| + L_M < \\ &< (L_A - L_B)|\mathbf{z}_J| + L_M \end{aligned} \qquad (B.32)$$

which implies that $|\mathbf{z}_j| < |\mathbf{z}_J|$. Since J cannot be chosen, it has to be reset by the *vigilance subsystem*, which means that $|\mathbf{I} \cap \mathbf{z}_J| < \rho|\mathbf{I}|$. Therefore,

$$L_A|\mathbf{z}_j| - L_B|\mathbf{z}_j| < L_A|\mathbf{I} \cap \mathbf{z}_J| - L_B|\mathbf{z}_J| < L_A\rho|\mathbf{I}| - L_B|\mathbf{z}_J| <$$
$$< L_A\rho|\mathbf{I}| - L_B|\mathbf{z}_j| \qquad (B.33)$$

which implies that,

$$|\mathbf{I} \cap \mathbf{z}_j| = |\mathbf{z}_j| < \rho|\mathbf{I}| \qquad (B.34)$$

B.8 SEARCH ORDER(THEOREM 7 OF ORIGINAL ART1)

The original Theorem 7 (page 100 of [Carpenter, 1987]) states the following:

Suppose that input pattern satisfies

$$L - 1 \leq \frac{1}{|\mathbf{I}|} \qquad (B.35)$$

and

$$|\mathbf{I}| \leq N - 1 \qquad (B.36)$$

Then $F2$ nodes are searched in the following order, if they are searched at all.

Subset templates with respect to **I** are searched first, in order of decreasing size. If the largest subset template is reset, then all subset templates are reset. If all subset templates have been reset and if no other learned templates exist, then the first uncommitted node to be activated will code **I**. If all subset templates are searched and if there exist learned superset templates but no mixed templates, then the node with the smallest superset template will be activated next and will code **I**. If all subset templates are searched and if both superset templates $\mathbf{z}_J$ and mixed templates $\mathbf{z}_j$ exist, then j will be searched before J if and only if

$$|\mathbf{z}_j| < |\mathbf{z}_J| \quad and \quad \frac{|\mathbf{I}|}{|\mathbf{z}_J|} < \frac{|\mathbf{I} \cap \mathbf{z}_j|}{|\mathbf{z}_j|} \qquad (B.37)$$

If all subset templates are searched and if there exist mixed templates but no superset templates, then a node j with a mixed template will be searched before an uncommitted node J if and only if

$$\frac{L|\mathbf{I} \cap \mathbf{z}_j|}{L - 1 + |\mathbf{z}_j|} > T_J(\mathbf{I}, t = 0). \qquad (B.38)$$

APPENDIX B: COMPUTATIONAL EQUIVALENCE OF ART1 AND ART1M   203

Where $T_J(\mathbf{I}, t = 0) = (L \sum I_i z_{iJ}(0))/(L - 1 + \sum z_{iJ}(0))$. The conditions expressed in eqs. (B.35)-(B.38) have to be changed in order to adapt this theorem to the modified ART1 architecture. The original proof will not be reproduced here, because it differs drastically from the one we will provide for the modified theorem. The modified theorem is identical to the original one, except for eqs. (B.35)-(B.38). It states the following:

Suppose that

$$\frac{L_A}{L_B} < \frac{N}{N-1} \tag{B.39}$$

and input pattern satisfies

$$|\mathbf{I}| \leq N - 1 \tag{B.40}$$

Then $F2$ nodes are searched in the following order, if they are searched at all.

Subset templates with respect to $\mathbf{I}$ are searched first, in order of decreasing size. If the largest subset template is reset, then all subset templates are reset. If all subset templates have been reset and if no other learned templates exist, then the first uncommitted node to be activated will code $\mathbf{I}$. If all subset templates are searched and if there exist learned superset templates but no mixed templates, then the node with the smallest superset template will be activated next and will code $\mathbf{I}$. If all subset templates are searched and if both superset templates $\mathbf{z}_J$ and mixed templates $\mathbf{z}_j$ exist, then j will be searched before J if and only if

$$|\mathbf{z}_j| < |\mathbf{z}_J| \quad \text{and} \quad \frac{|\mathbf{I}| - |\mathbf{I} \cap \mathbf{z}_j|}{|\mathbf{z}_J| - |\mathbf{z}_j|} < \frac{L_B}{L_A} \tag{B.41}$$

If all subset templates are searched and if there exist mixed templates but no superset templates, then a node j with a mixed template will be searched before an uncommitted node J if and only if

$$L_A |\mathbf{I} \cap \mathbf{z}_j| - L_B |\mathbf{z}_j| + L_M > T_J(\mathbf{I}, t = 0). \tag{B.42}$$

Where $T_J(\mathbf{I}, t = 0) = L_A \sum I_i z_{iJ}(0) - L_B \sum z_{iJ}(0) + L_M$. The proof has several parts:

1. First we show that a node J with a subset template $(\mathbf{I} \cap \mathbf{z}_J = \mathbf{z}_J)$ is searched before any node j with a non subset template. In this case,

$$O_j = L_A|\mathbf{I} \cap \mathbf{z}_j| - L_B|\mathbf{z}_j| + L_M = \quad (B.43)$$
$$= |\mathbf{I} \cap \mathbf{z}_j|(L_A - L_B \frac{|\mathbf{z}_j|}{|\mathbf{I} \cap \mathbf{z}_j|}) + L_M$$

Now, note that

$$\frac{|\mathbf{z}_j|}{|\mathbf{I} \cap \mathbf{z}_j|} > \frac{N}{N-1} \quad (B.44)$$

because[4]

$$\frac{|\mathbf{z}_j|}{|\mathbf{I} \cap \mathbf{z}_j|}|_{min} = \frac{|\mathbf{z}_j|}{|\mathbf{z}_j|-1}|_{min} = \frac{N-1}{N-2} > \frac{N}{N-1} \quad (B.45)$$

From eqs. (2.4), (B.39) and (B.44), it follows that

$$O_j < |\mathbf{I} \cap \mathbf{z}_j| L_B (\frac{L_A}{L_B} - \frac{N}{N-1}) + L_M < L_M \quad (B.46)$$

On the other hand,

$$O_J = (L_A - L_B)|\mathbf{z}_J| + L_M > L_M \quad (B.47)$$

Therefore,

$$O_J > O_j \quad (B.48)$$

2. Subset templates are searched in order of decreasing size:
Suppose two subset templates of $\mathbf{I}$, $\mathbf{z}_J$ and $\mathbf{z}_j$ such that $|\mathbf{z}_J| > |\mathbf{z}_j|$. Then

$$O_J = (L_A - L_B)|\mathbf{z}_J| + L_M > (L_A - L_B)|\mathbf{z}_j| + L_M = O_j \quad (B.49)$$

Therefore node J will be searched before node j. By eq. (B.31), if the largest subset template is reset, then all other subset templates are reset as well.

---

[4] We are assuming that j is not an uncommitted node ($|\mathbf{z}_j| < N$).

3. Subset templates J are searched before an uncommitted node j:

$$\begin{aligned} O_j &= L_A|\mathbf{I}| - L_B N + L_M \leq L_A(N-1) - L_B N + L_M = \\ &= L_B(\frac{L_A}{L_B}(N-1) - N) + L_M < L_B(\frac{N}{N-1}(N-1) - N) + L_M = \\ &= L_M < (L_A - L_B)|\mathbf{z}_J| + L_M = O_J \end{aligned} \quad (B.50)$$

Therefore now, if all subset templates are searched and if no other learned template exists, then an uncommitted node will be activated and code the input pattern.

4. If all subset templates have been searched and there exist learned superset templates but no mixed templates, the node with the smallest superset template J will be activated (and not an uncommitted node j) and code $\mathbf{I}$:

$$O_J = L_A|\mathbf{I}| - L_B|\mathbf{z}_J| + L_M > L_A|\mathbf{I}| - L_B N + L_M = O_j \quad (B.51)$$

If there are more than one superset templates, the one with the smallest $|\mathbf{z}_J|$ will be activated. Since $|\mathbf{I} \cap \mathbf{z}_J| = |\mathbf{I}| \geq \rho|\mathbf{I}|$ there is no reset, and $\mathbf{I}$ will be coded.

5. If all subset templates have been searched and there exist a superset template J and a mixed template j, then $O_j > O_J$ if and only if eq. (B.41) holds:

$$O_j - O_J = L_A(|\mathbf{I} \cap \mathbf{z}_j| - |\mathbf{I}|) + L_B(|\mathbf{z}_J| - |\mathbf{z}_j|) \quad (B.52)$$

(a) if eq. (B.41) holds:

$$O_j - O_J = L_A(\frac{L_B}{L_A} - \frac{|\mathbf{I}| - |\mathbf{I} \cap \mathbf{z}_j|}{|\mathbf{z}_J| - |\mathbf{z}_j|})(|\mathbf{z}_J| - |\mathbf{z}_j|) > 0 \quad (B.53)$$

(b) if $O_j > O_J$:
Assume first that $|\mathbf{z}_J| - |\mathbf{z}_j| < 0$. Then, by eq. (B.53), it has to be

$$\frac{L_B}{L_A} < \frac{|\mathbf{I}| - |\mathbf{I} \cap \mathbf{z}_j|}{|\mathbf{z}_J| - |\mathbf{z}_j|} \quad (B.54)$$

Since $L_A > L_B > 0$ it had to be $|\mathbf{I}| - |\mathbf{I} \cap \mathbf{z}_j| < 0$, which is false. Therefore, it must be $|\mathbf{z}_J| - |\mathbf{z}_j| > 0$ and

$$\frac{L_B}{L_A} > \frac{|\mathbf{I}| - |\mathbf{I} \cap \mathbf{z}_j|}{|\mathbf{z}_J| - |\mathbf{z}_j|} \quad (B.55)$$

6. If all subset templates are searched and if there exist mixed templates but no superset templates, then a node j with a mixed template ($O_j = L_A|\mathbf{I} \cap \mathbf{z}_j| - L_B|\mathbf{z}_j| + L_M$) will be searched before an uncommitted node J ($O_J = L_A|\mathbf{I}| - L_B N + L_M$) if and only if eq. (B.42) holds:

$$O_j - O_J = L_A(|\mathbf{I} \cap \mathbf{z}_j| - |\mathbf{I}|) - L_B(|\mathbf{z}_j| - N) > 0 \Leftrightarrow \quad (B.56)$$
$$\Leftrightarrow L_A|\mathbf{I} \cap \mathbf{z}_j| - L_B|\mathbf{z}_j| + L_M > L_A|\mathbf{I}| - L_B N + L_M = T_J(\mathbf{I}, t=0)$$

This completes the proof of the modified Theorem 7 for the modified ART1 architecture.

B.9 BIASING THE NETWORK TOWARDS UNCOMMITTED NODES

In the original ART1 architecture, choosing L large increases the tendency of the network to choose uncommitted nodes in response to unfamiliar input patterns $\mathbf{I}$. In the modified ART1 architecture, the same effect is observed when choosing $\alpha = L_A/L_B$ large. This can be understood through the following reasoning.

When an input pattern $\mathbf{I}$ is presented, an uncommitted node is chosen before a coded node j if

$$L_A|\mathbf{I} \cap \mathbf{z}_j| - L_B|\mathbf{z}_j| < L_A|\mathbf{I}| - L_B N \quad (B.57)$$

This inequality is equivalent to

$$\frac{L_A}{L_B} > \frac{N - |\mathbf{z}_j|}{|\mathbf{I}| - |\mathbf{I} \cap \mathbf{z}_j|} \quad (B.58)$$

As the ratio $\alpha = L_A/L_B$ increases it is more likely that eq. (B.58) is satisfied, and hence that uncommitted nodes are chosen before coded nodes, regardless of the *vigilance parameter* value ρ.

B.10 EXPANDING PROOFS TO FUZZY–ART

All properties, theorems and proofs in this Appendix are directly applicable to Fuzzy–ART by simply substituting the intersection operator ($\cap$) by the fuzzy minimum operator ($\wedge$). Note that the subset and superset concepts used in ART1 also expand to the fuzzy subset and superset concepts in Fuzzy–ART by doing this simple substitution. In ART1 pattern $\mathbf{a}$ is said to be a subset of pattern $\mathbf{b}$ (or $\mathbf{b}$ a superset of $\mathbf{a}$), and is denoted as $\mathbf{a} \subset \mathbf{b}$, if

$$\mathbf{a} \cap \mathbf{b} = \mathbf{a} \quad (B.59)$$

In Fuzzy–ART pattern **a** is said to be a fuzzy subset of pattern **b** (or **b** a fuzzy superset of **a**), and is also denoted as **a** $\subset$ **b**, if [Zadeh, 1965]

$$\mathbf{a} \wedge \mathbf{b} = \mathbf{a} \tag{B.60}$$

B.11 REMARKS

Even though in this Appendix we have shown that the computational properties of the original ART1 system are preserved in the modified ART1 system, the response of both systems to an arbitrary list of training patterns will not be exactly the same. The main underlying reason for this difference in behavior is that the initial ordering

$$O_{j_1} > O_{j_2} > O_{j_3} > \ldots \tag{B.61}$$

is not always exactly the same for both architectures. In Chapter 2 we tried to study the differences in behavior between the two ART1 systems. As we saw, for most cases the behavior is identical, although in a few cases a different behavior results.

Appendix C
Systematic Width-and-Length Dependent CMOS Transistor Mismatch Characterization

Precise analog CMOS circuit design requires availability of confident transistor mismatch models during the design and simulation stages. During the design phase of an analog VLSI circuit, designers face many constraints imposed by the design specifications, such as speed, bandwidth, noise, precision, power consumption, area consumption, which need to be traded off for optimum overall performance. Designers must rely on accurate simulation tools in order to achieve a well optimized final design, specially if performance is pushed to the limits allowed by a given fabrication process. Simulation tools are reliable as long as they are based on good models and confident characterization techniques. If good and well characterized models are embedded in a reliable simulator, circuit designers can confidently test different circuit topologies and optimize each one of them by optimally sizing their transistors. Automatic design tools are available that by interacting with a simulator are able to obtain transistor sizes for close-to-optimum performance for a given circuit topology and a set of design constraints [Medeiro, 1994].

Many times it is not possible to simulate properly the precision limits that can be achieved by a certain circuit topology in a given fabrication process because VLSI circuit manufacturers rarely provide transistor mismatch information, and, if they do, its dependence on transistor size (width and length, independently[1]) is not known. In this Appendix we provide a very simple and cheap methodology to characterize transistor mismatch as a function of transistor width and length, and how to use this information to predict mismatch effects in circuit simulators.

---

[1] Sometimes manufactures provide mismatch information as a function of transistor area, but this information has been obtained for (almost) square transistors [Pelgrom, 1989].

210   ADAPTIVE RESONANCE THEORY MICROCHIPS

In the specialized literature transistor mismatch is usually characterized by providing the standard deviation of a set of transistor electrical parameters such as the threshold voltage V_{TO}, the current gain factor $\beta = \mu C_{ox} W/L$ (μ is mobility, C_{ox} is gate oxide capacitance density, W is transistor width, and L is transistor length), and bulk threshold parameter γ. Table C.1 shows a few examples [Pelgrom, 1989], [Lakshmikumar, 1986], [Bastos, 1995], [Bastos, 1997] on what dependencies of $\sigma^2_{(\Delta\beta/\beta)}$, $\sigma^2_{(\Delta V_{TO})}$, and $\sigma^2_{(\Delta\gamma)}$ on transistor sizes ($x = 1/W$, $y = 1/L$, W is transistor width, L is transistor length) and distance D have been postulated. A very nice study [Michael, 1992] based on BSIM transistor models is also available in the literature.

In the present paper we provide an experimental method to obtain a relatively high number of samples (of $\sigma^2_{(\Delta\beta/\beta)}$, $\sigma^2_{(\Delta V_{TO})}$, $\sigma^2_{(\Delta\gamma)}$ and) in the $\{x,y\}$ space. Then we will fit these measured samples to a function

$$\sigma^2_{(\Delta P)} = C_{00} + C_{10}x + C_{01}y + C_{20}x^2 + C_{11}xy + C_{02}y^2 + \cdots =$$
$$= \sum_{n,m} C_{nm} x^n y^m \qquad (C.1)$$

where ΔP is the observed mismatch of a certain electrical parameter (like $\Delta\beta/\beta$, ΔV_{TO}, or $\Delta\gamma$). Note that we are not interested in discovering the physical meaning of coefficients C_{nm}, but just in obtaining a good approximation for the function $\sigma^2_{(\Delta P)} = f(x,y)$ in order to use it confidently in a circuit simulator. Note also that the $\{x,y\}$ space limits are $x_{max} = 1/W_{min}$, $y_{max} = 1/L_{min}$, $x_{min} = 0$, $y_{min} = 0$. Measuring a reasonable high number of sample points in this $\{x,y\}$ space provides sufficient information to interpolate the functions $\sigma^2_{(\Delta P)}$, which are fairly smooth in this space. Next Section describes the mismatch characterization chip used to obtain all characterization data.

C.1 MISMATCH CHARACTERIZATION CHIP

According to Table C.1 the mismatch in parameter P between two identical transistors is statistically characterized by a quadratic deviation whose general form is

$$\sigma^2_{(\Delta P)} = f(x,y) + S_P^2 D^2 \qquad (C.2)$$

where $x = 1/W$, $y = 1/L$, W and L are the transistor width and length, and D is the distance between them. The two terms in eq. (C.2) indicate that there are two physical causes producing transistor mismatch. The first term is produced by the fact that **device physical parameters** (like doping concentrations, junctions depth, implants depth, oxide thicknesses, ...) are not

APPENDIX C: CMOS TRANSISTOR MISMATCH CHARACTERIZATION

Table C.1. Examples of Mismatch Models in the Literature ($x = 1/W$, $y = 1/L$)

	$\sigma^2_{(\Delta\beta/\beta)}$	$\sigma^2_{(\Delta V_{TO})}$	$\sigma^2_{(\Delta\gamma)}$
[Pelgrom, 1989]	$A_s^2 xy + A_w^2 x^2 y + \\ + A_L^2 xy^2 + S_\beta^2 D^2$	$A_{V_{TO}}^2 xy + S_{V_{TO}}^2 D^2$	$A_\gamma^2 xy + \\ + S_\gamma^2 D^2$
[Lakshmikumar, 1986]	$A_{\beta 1} xy + \\ + A_{\beta 2}(x^2 + y^2)$	$A_{VO} xy$	-
[Bastos, 1995]	$A_\beta^2 xy$	$A_{V1}^2 xy + A_{V2}^2 xy^2 - \\ - A_{V3}^2 x^2 y$	-

exactly constant but suffer from noise-like perturbations along a die. By increasing transistor areas the **device electrical parameters** P (like threshold voltage V_{TO}, current gain factor β, or bulk threshold parameter γ) will become less sensitive to the noisy nature of the **device physical parameters**. The second term in eq. (C.2), characterized by parameter S_P, is produced by the fact that the **device physical parameters** present a certain gradient variation along the dies. Usually, the gradients present in small and medium size dies can be approximated by planes. Statistical characterization of these planes (which means obtaining S_P) can be performed with a small number of transistors per die and measuring many dies. On the other hand, the transistor mismatch induced by the **device physical parameters** noisy nature, changes little from die to die. Consequently, its statistical characterization can be done by putting many transistors in a single die and measuring a reduced number of dies. This is very convenient for a 'do-it-yourself' working style, since circuit designers can easily have a small number of samples of a prototype chip at a reasonable cost. This Appendix thus concentrates on the characterization of size dependent mismatch terms, and a wide range of transistor sizes will be characterized. On the contrary, note that characterization of distance terms (like S_P) is less critical for circuit designers because gradient-induced mismatches can be compensated through layout techniques (like common centroids, for example).

With all this in mind we designed a special purpose chip intended to characterize the 'noise- induced-terms' of CMOS transistor mismatches, as a function of transistor size. As shown in Fig. C.1, the chip consists of an array of identical cells. Each cell contains 30 NMOS and 30 PMOS transistors, each of a different size. Sizes are such that widths are $W = 40\mu m$, $20\mu m$, $10\mu m$, $5\mu m$, $2.5\mu m$, $1.25\mu m$, and lengths are $L = 40\mu m$, $10\mu m$, $4\mu m$, $2\mu m$, $1\mu m$. Digital decoding/selection circuitry is included in each cell and outside the array. Elements in the chip are arranged in a way such that all NMOS transistors have their Drains connected to chip pin DN, all PMOS transistors have their

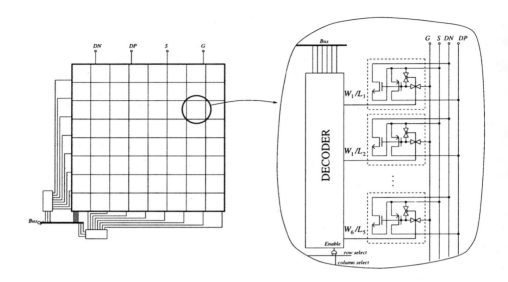

Figure C.1. Mismatch Characterization Chip Simplified Schematic

Drains connected to chip pin DP, all NMOS and PMOS transistors have their sources connected to chip pin S, all NMOS and PMOS transistors have their Gates short-circuited to their sources, except for one NMOS-PMOS pair which has their Gates connected to chip pin G. The digital bus and the internal decoding/selection circuitry selects one cell in the array and, inside this cell, one pair of NMOS and PMOS transistors, connecting their Gates to chip pin G. A chip with an 8×8 cell array has been fabricated in a digital $1.0 \mu m$ CMOS process which occupies an area of $4.0mm \times 3.5mm$, and uses 18 pins (12 for the decoding/selection Bus, DN, DP, G, S, V_{dd}, and Gnd). Some transistors in the periphery cells presented large systematic deviations with respect to those in the inside cells. Consequently, statistical computations were performed only on inner cells transistors, thus rendering an effective cell array of 6×6.

The experimental characterization set-up consists of a host computer controlling the decoding/selection bus and a DC curves measuring instrument (like the HP4145). This instrument is connected to pins DN, DP, S, G, and chip substrate. The host computer selects one NMOS-PMOS pair and the instrument measures first the NMOS transistor (putting connection DP into high-impedance and measuring through pins S, G, and DN) and then the PMOS transistor (putting connection DN into high-impedance and measuring through pins S, G, and DP). A simple software program sequentially selects and measures all transistors in the chip. Next Section describes the DC curves that were measured for each transistor, how electrical parameter mismatches were extracted from these curves, and how their statistical characterization was performed.

C.2 MISMATCH PARAMETER EXTRACTION AND STATISTICAL CHARACTERIZATION

Transistor parameter mismatches were obtained by measuring pairs of identical transistors located in adjacent cells of the same rows. Since in the chip there are 6 × 6 effective cells, there are 6 rows, each of which provides 5 pairs of adjacent cells. This results in 30 adjacent transistor pairs (for each transistor size and type). The statistical significance of 30 measurements to determine a standard deviation is as follows: assuming a normal distribution, if 30 samples are available to compute a standard deviation $\sigma_{Computed}$, it can be assured that the 95% confidence interval for the real standard deviation σ_{Real} is [Rade, 1990]

$$0.7964 \times \sigma_{Computed} \leq \sigma_{Real} \leq 1.344 \times \sigma_{Computed}. \quad (C.3)$$

For each transistor pair, two curves were measured while operating in the ohmic region (always in strong inversion). These curves are[2]

Curve 1: $\quad I_{DS}(V_{GS}), V_{SB} = 0V,$ $\quad$ (C.4)
$V_{DS} = 0.1V, V_{GS} \in [V_{GS_{min}}, V_{GS_{max}}]$

Curve 2: $\quad I_{DS}(V_{SB}), V_{GS} = 3.0V,$ $\quad$ (C.5)
$V_{DS} = 0.1V, V_{SB} \in [V_{SB_{min}}, V_{SB_{max}}]$

Care must be taken in order to keep current levels sufficiently small so that mismatch introduced by series resistances (contact resistances, variable length routing wires, ...) is negligible. The following strong inversion ohmic region transistor model was assumed,

$$I_{DS} = \beta \frac{V_{GS} - V_T(V_{SB}) - \frac{1}{2}V_{DS}}{1 + \theta(V_{GS} - V_T(V_{SB}))} V_{DS} \quad (C.6)$$

$$V_T(V_{SB}) = V_{TO} + \gamma \left[\sqrt{\Phi + V_{SB}} - \sqrt{\Phi}\right] \quad (C.7)$$

---

[2] All voltages and currents are taken in absolute value, so that the same expressions are valid for NMOS and PMOS transistors.

which renders the following current mismatch for each transistor pair:

$$\frac{\Delta I_{DS}}{I_{DS}} = \frac{\Delta \beta}{\beta} - \frac{1+\frac{1}{2}\theta V_{DS}}{V_{GS}-V_T(V_{SB})-\frac{1}{2}V_{DS}}\Delta V_T - \qquad (C.8)$$
$$- \frac{V_{GS}-V_T(V_{SB})}{1+\theta(V_{GS}-V_T(V_{SB}))}\Delta\theta$$

$$\Delta V_T = \Delta V_{TO} + \Delta\gamma\left[\sqrt{\Phi+V_{SB}}-\sqrt{\Phi}\right] \qquad (C.9)$$

The drain and source series resistances (due to contacts, diffusion resistance, and metal routing lines) have the effect of changing the extracted value of the mobility reduction parameter θ in eq. (C.6) [Pelgrom, 1989],

$$\theta = \theta_{Real} + \beta R_{DS} \qquad (C.10)$$

where θ_{Real} is the real mobility reduction parameter of the transistor and R_{DS} is the sum of the series resistances at drain and source. The mismatch contribution of $\Delta(\beta R_{DS})$ can be of the order or higher than that of $\Delta\theta_{Real}$, but both contribute very little to $\Delta I_{DS}/I_{DS}$.

For each transistor pair, the measurement/extraction procedure was as follows:

- **Curve 1** (eq. (C.4)) was measured for both transistors. Using the Levenberg-Marquardt nonlinear curve fitting technique, the first curve was fitted to eq. (C.6) and parameters β, V_{TO}, and θ were obtained. Using the two measured curves, the curve $\Delta I_{DS}/I_{DS}$ was computed and fitted to eq. (C.8) obtaining $\Delta\beta/\beta$, ΔV_{TO}, and $\Delta\theta$ for this transistor pair.

- **Curve 2** (eq. (C.5)) was measured for both transistors, and curve $\Delta I_{DS}/I_{DS}$ was computed. According to eqs. (C.8), (C.9) it must fit to

$$\frac{\Delta I_{DS}}{I_{DS}} = \frac{\Delta\beta}{\beta} - \frac{1+\frac{1}{2}\theta V_{DS}}{V_x - \frac{1}{2}V_{DS}}\left[\Delta V_{TO}+\Delta\gamma(\sqrt{\Phi+V_{SB}}-\sqrt{\Phi})\right] -$$
$$- \frac{V_x}{1+\theta V_x}\Delta\theta, \qquad (C.11)$$

where $V_x = V_{GS}-V_T(V_{SB})$. For this transistor pair $\Delta\beta/\beta$ and ΔV_{TO} are already known. The values for $V_x = V_{GS}-V_T(V_{SB})$ can be obtained from eq. (C.6)

$$V_x = \frac{I_{DS} + \frac{\beta}{2}V_{DS}^2}{\beta V_{DS} - \theta I_{DS}} \tag{C.12}$$

since β, θ, and V_{DS} are already known, and the I_{DS} values are those just measured at Curve 2. Consequently, $\Delta\gamma$ is the only unknown parameter in eq. (C.11), which can be obtained for this pair by fitting eq. (C.11), after substituting eq. (C.12) into it.

This measurement/extraction procedure is repeated for the $N_T = 30$ transistor pairs. For each extracted mismatch parameter ΔP ($\Delta\beta/\beta$, ΔV_{TO}, $\Delta\gamma$) its standard deviation

$$\sigma_{(\Delta P)}^2 = \frac{1}{N_T}\sum_{n=1}^{N_T}(\Delta P_n - \overline{\Delta P})^2 \tag{C.13}$$

is computed. For each fabricated chip, eq. (C.13) should be obtained for each transistor size and type (NMOS or PMOS).

C.3 CHARACTERIZATION RESULTS

A mismatch characterization chip was fabricated in a digital double-metal single-poly $1.0\mu m$ CMOS process. The die area of the chip is $3.5mm \times 4.0mm$. Ten samples were delivered by the foundry, eight of which were fault free. For each die, transistor size, and transistor type the following parameters were extracted, following the procedure described in the previous Section,

$$\sigma_{(\Delta\beta/\beta)} \quad , \quad \sigma_{(\Delta V_{TO})} \quad , \quad \sigma_{(\Delta\gamma)} \tag{C.14}$$

Table C.2 shows these parameters for the NMOS transistors, averaged over all dies and indicating the spread from die to die. Table C.3 shows the average over all dies of these parameters for the PMOS transistors. Each cell in Tables C.2 and C.3 indicates

$$\overline{\sigma_{(\Delta P)}} \pm 3\sigma(\sigma_{(\Delta P)}) \tag{C.15}$$

where,

$$\overline{\sigma_{(\Delta P)}} = \frac{1}{N_{Dies}} \sum_{n_d=1}^{N_{Dies}} \sigma_{(\Delta P)}(n_d) \tag{C.16}$$

$$\sigma(\sigma_{(\Delta P)}) = \frac{1}{N_{Dies}} \sum_{n_d=1}^{N_{Dies}} (\sigma_{(\Delta P)}(n_d) - \overline{\sigma_{(\Delta P)}})^2$$

and N_{Dies} is the total number of fault-free dies.

Looking at the deviations $\sigma_{(\Delta P)}$ in Tables C.2 and C.3 and at their $\pm 3\sigma$ inter-chip spread, one can see that the 3σ spread is of the order of $\pm 50\%$ of the average deviation. This shows that the deviations $\sigma_{(\Delta P)}$ of transistor mismatch parameters have a fairly stable behavior from chip to chip.

In order to measure the precision of the extracted deviation values, one of the dies was measured without sweeping the transistor pairs: Curves 1 and 2 (eqs. (C.4) and (C.5)) were measured 30 times for the same pair. The resulting deviations indicate the measurement set-up and extraction procedure precision. This is given in Table C.4 for the NMOS transistors and in Table C.5 for the PMOS transistors. Units in Table C.4 and C.5 are the same than for Tables C.2 and C.3, except for a 10^{-3} factor.

The extracted data for each column in Tables C.2 and C.3 can be considered to be sample points of a two dimensional surface whose independent variables are $x = 1/W$ and $y = 1/L$,

$$\sigma_{(\Delta P)} = f(x, y) \tag{C.17}$$

As mentioned before, several functionals have been attributed to eq. (C.17). We will assume the following polynomial dependency

$$\begin{aligned} \sigma^2_{(\Delta P)_{fit}}(x, y, C_{nm}) &= C_{00} + C_{10}x + C_{01}y + C_{20}x^2 + C_{11}xy + C_{02}y^2 + \cdots = \\ &= \sum_{n,m} C_{nm} x^n y^m \end{aligned} \tag{C.18}$$

The optimum set of coefficients C_{nm} were obtained by fitting eq. (C.18) (using Least Mean Squares) to the experimental data $\sigma_{(\Delta P)}(x_i, y_i, n_d)$ for all sizes and dies. For this, the following error term was defined,

$$\epsilon = \sum_{n_d=1}^{N_{Dies}} \sum_{i=1}^{N_{Sizes}} w_i \left[\sigma^2_{(\Delta P)}(x_i, y_i, n_d) - \sigma^2_{(\Delta P)_{fit}}(x_i, y_i, C_{nm}) \right]^2 \tag{C.19}$$

Table C.2. NMOS Mismatch Characterization Parameters averaged over all measured dies and indicating the $\pm 3\sigma$ spread from die to die

sizes	$\sigma_{(\Delta\beta/\beta)}(\times 10^{-3})$	$\sigma_{(\Delta V_{TO})}(mV)$	$\sigma_{(\Delta\gamma)}(mV^{1/2})$
40/40	1.18 ± 0.78	0.60 ± 0.35	0.49 ± 0.15
40/10	2.21 ± 1.71	1.21 ± 0.59	0.57 ± 0.22
40/4	3.36 ± 1.71	1.27 ± 0.52	0.86 ± 0.36
40/2	5.92 ± 1.41	1.94 ± 0.76	1.29 ± 0.71
40/1	9.01 ± 2.93	5.23 ± 1.97	3.93 ± 1.28
20/40	1.87 ± 0.66	0.66 ± 0.25	0.48 ± 0.27
20/10	2.43 ± 0.90	1.14 ± 0.40	0.73 ± 0.34
20/4	4.17 ± 1.75	1.67 ± 0.56	1.14 ± 0.46
20/2	9.18 ± 4.74	2.58 ± 1.84	1.80 ± 1.07
20/1	12.52 ± 7.10	6.67 ± 1.95	4.72 ± 1.19
10/40	1.77 ± 0.51	0.66 ± 0.24	0.62 ± 0.49
10/10	4.87 ± 1.45	1.37 ± 0.41	0.92 ± 0.43
10/4	6.83 ± 1.94	2.03 ± 0.88	1.39 ± 0.37
10/2	7.89 ± 2.23	3.70 ± 1.69	2.27 ± 1.29
10/1	13.41 ± 7.36	8.60 ± 2.30	5.49 ± 3.23
5/40	2.97 ± 1.26	0.82 ± 0.61	0.76 ± 0.25
5/10	4.53 ± 2.28	2.01 ± 0.68	1.35 ± 0.56
5/4	6.77 ± 1.90	3.29 ± 1.04	2.02 ± 1.24
5/2	10.10 ± 4.28	4.76 ± 2.21	3.13 ± 1.14
5/1	14.80 ± 4.00	11.66 ± 3.10	6.15 ± 2.54
2.5/40	6.71 ± 2.88	1.27 ± 0.86	1.01 ± 0.40
2.5/10	8.84 ± 5.01	2.46 ± 1.26	1.66 ± 0.78
2.5/4	9.09 ± 3.69	4.14 ± 1.43	2.98 ± 1.50
2.5/2	13.85 ± 5.46	6.68 ± 3.32	4.30 ± 1.05
2.5/1	21.56 ± 10.77	15.93 ± 8.49	8.51 ± 3.28
1.25/40	11.40 ± 4.56	1.67 ± 0.87	1.79 ± 0.56
1.25/10	12.50 ± 4.96	3.52 ± 1.92	2.69 ± 1.34
1.25/4	10.71 ± 2.58	5.24 ± 2.06	3.75 ± 1.53
1.25/2	15.26 ± 8.25	9.65 ± 4.65	5.69 ± 2.66
1.25/1	21.69 ± 9.06	21.94 ± 6.02	11.06 ± 4.29

where w_i is a weighting term that depends on the spread of $\sigma_{(\Delta P)}(x_i, y_i, n_d)$ from die to die, and N_{Sizes} is the total number of transistor sizes available in the chip. If $\sigma_{(\Delta P)}$, for size $x_i = 1/W_i$, $y_i = 1/L_i$ has a large spread from die to die then weight w_i will be smaller than for another size whose spread is smaller. If for a given size (x_i, y_i) the spread from die to die is $\sigma(\sigma_{(\Delta P)}(x_i, y_i))$ then the weight w_i is defined as

Table C.3. PMOS Mismatch Characterization Parameters averaged over all measured dies and indicating the $\pm 3\sigma$ spread from die to die

sizes	$\sigma_{(\Delta\beta/\beta)}(\times 10^{-3})$	$\sigma_{(\Delta V_{TO})}(mV)$	$\sigma_{(\Delta\gamma)}(mV^{1/2})$
40/40	1.41 ± 0.67	1.38 ± 0.77	0.45 ± 0.18
40/10	2.20 ± 1.35	1.89 ± 1.54	0.53 ± 0.20
40/4	3.10 ± 1.41	1.87 ± 0.73	0.64 ± 0.16
40/2	6.22 ± 2.70	2.73 ± 0.69	1.08 ± 0.41
40/1	11.43 ± 6.39	5.78 ± 2.33	3.65 ± 1.70
20/40	1.75 ± 0.49	1.06 ± 0.87	0.54 ± 0.20
20/10	2.97 ± 1.20	1.80 ± 0.70	0.58 ± 0.19
20/4	3.77 ± 1.56	2.27 ± 1.09	0.83 ± 0.21
20/2	10.06 ± 4.41	3.56 ± 0.80	1.59 ± 0.89
20/1	13.55 ± 4.92	7.60 ± 2.61	3.81 ± 1.83
10/40	1.64 ± 0.74	1.19 ± 0.50	0.49 ± 0.22
10/10	4.76 ± 2.06	2.18 ± 1.10	0.73 ± 0.32
10/4	6.75 ± 2.03	2.78 ± 0.72	1.00 ± 0.52
10/2	9.03 ± 7.84	4.86 ± 3.14	1.93 ± 0.78
10/1	20.70 ± 6.41	10.66 ± 2.77	5.18 ± 1.83
5/40	2.89 ± 1.44	1.43 ± 0.56	0.55 ± 0.30
5/10	5.21 ± 1.60	2.54 ± 0.78	0.91 ± 0.51
5/4	6.86 ± 1.67	4.33 ± 1.90	1.35 ± 0.47
5/2	11.23 ± 7.76	7.56 ± 4.29	2.22 ± 0.52
5/1	17.99 ± 8.99	12.81 ± 4.41	4.98 ± 2.17
2.5/40	6.04 ± 1.94	1.73 ± 0.69	0.58 ± 0.23
2.5/10	9.44 ± 6.28	3.40 ± 0.77	1.12 ± 0.50
2.5/4	11.82 ± 4.01	5.86 ± 1.37	1.94 ± 0.82
2.5/2	14.39 ± 8.64	9.08 ± 4.18	3.14 ± 1.54
2.5/1	25.44 ± 8.31	17.63 ± 5.36	6.60 ± 2.40
1.25/40	12.22 ± 8.05	2.65 ± 1.73	0.95 ± 0.40
1.25/10	14.32 ± 7.29	4.50 ± 2.46	1.71 ± 0.70
1.25/4	12.46 ± 5.98	6.70 ± 3.00	2.49 ± 0.94
1.25/2	19.33 ± 10.56	10.76 ± 4.12	3.80 ± 1.49
1.25/1	28.54 ± 9.41	21.05 ± 7.97	7.18 ± 3.89

$$w_i = \frac{e^{-\Omega_i}}{\Omega_i^2} \quad , \quad \Omega_i = \frac{\sigma(\sigma_{(\Delta P)}(x_i, y_i))}{\overline{\sigma_{(\Delta P)}(x_i, y_i)}}. \tag{C.20}$$

Fig. C.2(a) shows the resulting fitted surface for $\sigma_{(\Delta\beta/\beta)_{fit}}(1/W, 1/L)$, for NMOS transistors. Also shown in Fig. C.2(a) (with diamonds) are the experimental measured values for $\sigma_{(\Delta\beta/\beta)}(1/W_i, 1/L_i, n_d)$ for each transistor size

Table C.4. NMOS Precision Measurements for the data in Table C.2

sizes	$\sigma_{(\Delta\beta/\beta)}$	$\sigma_{(\Delta V_{TO})}(V)$	$\sigma_{(\Delta\gamma)}(V^{1/2})$
40/40	2.99e-04	1.70e-04	2.64e-04
40/10	4.29e-04	2.54e-04	3.58e-04
40/4	3.84e-04	2.09e-04	4.36e-04
40/2	4.57e-04	2.88e-04	4.69e-04
40/1	6.50e-04	4.15e-04	6.62e-04
20/40	2.64e-04	1.40e-04	3.14e-04
20/10	2.80e-04	1.59e-04	3.29e-04
20/4	3.56e-04	1.92e-04	3.07e-04
20/2	4.35e-04	2.24e-04	4.98e-04
20/1	5.58e-04	3.92e-04	6.02e-04
10/40	3.92e-04	2.46e-04	3.10e-04
10/10	3.87e-04	2.39e-04	4.17e-04
10/4	4.51e-04	3.04e-04	4.00e-04
10/2	4.20e-04	2.44e-04	3.53e-04
10/1	6.11e-04	4.20e-04	7.70e-04
5/40	3.75e-04	1.71e-04	4.45e-04
5/10	4.38e-04	2.57e-04	3.71e-04
5/4	2.91e-04	1.80e-04	2.94e-04
5/2	3.49e-04	2.53e-04	3.76e-04
5/1	4.60e-04	2.47e-04	7.99e-04
2.5/40	2.80e-04	1.25e-04	3.68e-04
2.5/10	3.49e-04	2.32e-04	4.20e-04
2.5/4	3.64e-04	2.13e-04	4.77e-04
2.5/2	3.52e-04	2.07e-04	4.06e-04
2.5/1	4.20e-04	3.29e-04	4.97e-04
1.25/40	2.72e-04	1.76e-04	2.60e-04
1.25/10	3.54e-04	1.69e-04	3.24e-04
1.25/4	3.68e-04	2.27e-04	3.24e-04
1.25/2	4.20e-04	2.70e-04	4.53e-04
1.25/1	4.86e-04	2.55e-04	4.63e-04

and for each die, for NMOS transistors. Fig. C.2(b) shows the fitted surface and experimental data for the threshold voltage deviation $\sigma_{(\Delta V_{TO})}$, and Fig. C.2(c) does it for the bulk threshold parameter deviation $\sigma_{(\Delta\gamma)}$, for NMOS transistors. The coefficients C_{nm} that result from the fitting procedure are given in Tables C.6 and C.7 for all deviations for NMOS and PMOS transistors, respectively. The units of the coefficients are such that if x and y are expressed in μm^{-1} then the deviations σ have the units used in Tables C.2, C.3 (without the 10^{-3} factor), and Tables C.4, C.5.

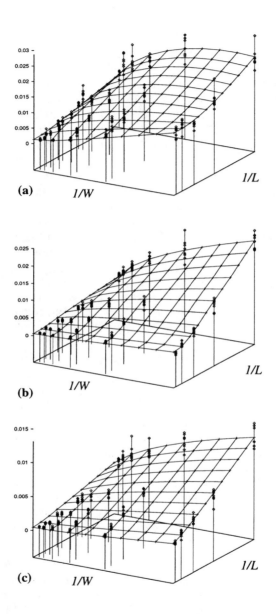

Figure C.2. Experimentally measured/extracted mismatch data (diamonds) as a function of transistor size, for NMOS transistors. Also shown are the interpolated surfaces. (a) Results for $\sigma_{(\Delta\beta/\beta)}$, (b) for $\sigma_{(\Delta V_{TO})}$, (c) and for $\sigma_{(\Delta\gamma)}$

APPENDIX C: CMOS TRANSISTOR MISMATCH CHARACTERIZATION   221

Table C.5. PMOS Precision Measurements for the data in Table C.3

sizes	$\sigma_{(\Delta\beta/\beta)}$	$\sigma_{(\Delta V_{TO})}(V)$	$\sigma_{(\Delta\gamma)}(V^{1/2})$
40/40	4.99e-04	2.95e-04	4.10e-04
40/10	4.01e-04	1.77e-04	4.70e-04
40/4	3.70e-04	1.60e-04	3.73e-04
40/2	4.23e-04	1.56e-04	4.06e-04
40/1	6.46e-04	2.67e-04	6.66e-04
20/40	4.67e-04	1.89e-04	3.70e-04
20/10	4.57e-04	2.24e-04	3.74e-04
20/4	3.77e-04	1.21e-04	4.60e-04
20/2	3.70e-04	1.98e-04	3.83e-04
20/1	5.53e-04	2.53e-04	4.98e-04
10/40	4.80e-04	1.97e-04	3.78e-04
10/10	4.22e-04	2.37e-04	3.86e-04
10/4	4.64e-04	1.80e-04	4.03e-04
10/2	5.29e-04	2.01e-04	4.74e-04
10/1	4.16e-04	2.33e-04	3.71e-04
5/40	3.43e-04	1.75e-04	3.51e-04
5/10	5.26e-04	1.74e-04	4.38e-04
5/4	4.02e-04	2.32e-04	3.76e-04
5/2	4.21e-04	2.05e-04	3.57e-04
5/1	4.85e-04	2.64e-04	4.81e-04
2.5/40	4.52e-04	2.81e-04	3.45e-04
2.5/10	4.43e-04	1.60e-04	3.91e-04
2.5/4	3.63e-04	1.26e-04	3.22e-04
2.5/2	3.40e-04	1.06e-04	4.04e-04
2.5/1	4.02e-04	2.37e-04	5.01e-04
1.25/40	4.26e-04	2.50e-04	5.03e-04
1.25/10	2.05e-04	6.80e-05	2.76e-04
1.25/4	3.34e-04	1.08e-04	3.08e-04
1.25/2	4.84e-04	1.41e-04	4.87e-04
1.25/1	6.41e-04	3.23e-04	5.65e-04

Correlations among deviations of different parameters can also be easily obtained. However, according to our data, no definite conclusions can be made. We observed that for some dies there were high correlation coefficients for some parameter pairs, while for other dies the same correlation coefficient could change significantly (we observed changes from almost +1 correlation factor for one die to almost −1 for another die). If these measurements are correct this would mean that deviation correlations could change significantly from one area of the wafer to another. However, before making any conclusions about

Table C.6. NMOS Resulting coefficients for the fitting functions

	C_{00}	C_{11}	C_{20}	C_{02}	C_{21}	C_{12}	C_{22}
$\sigma^2_{(\Delta\beta/\beta)}$	1.7e-06	8.9e-04	2.0e-04	5.1e-05	-1.2e-03	4.2e-04	0
$\sigma^2_{(\Delta V_{TO})}$	3.4e-07	-4.3e-05	8.1e-06	1.6e-05	0	6.2e-04	0
$\sigma^2_{(\Delta\gamma)}$	2.0e-07	-7.5e-06	5.1e-06	1.3e-05	3.3e-05	1.6e-04	-5.7e-05

Table C.7. PMOS Resulting coefficients for the fitting functions

	C_{00}	C_{11}	C_{20}	C_{02}	C_{21}	C_{12}	C_{22}
$\sigma^2_{(\Delta\beta/\beta)}$	1.4e-06	7.1e-04	2.6e-04	1.5e-04	-6.0e-04	9.1e-04	-6.5e-04
$\sigma^2_{(\Delta V_{TO})}$	2.5e-06	1.7e-04	0	0	-1.3e-04	8.4e-04	-4.5e-04
$\sigma^2_{(\Delta\gamma)}$	1.7e-07	-2.0e-05	0	1.0e-05	5.5e-05	1.3e-04	-1.3e-04

this possible "wafer-level" behavior it would be necessary to make intensive measurements using many dies per wafer and for many wafers. In our case, since we only used 8 dies from one run we can only make approximate conclusions regarding the measured standard deviation values, but not much can be said about their correlations.

References

Allen, P. E. and Holberg, D. R. (1987). *CMOS Analog Design*. Holt Rinehart and Winston Inc., New York.

Andreou, A. G., Boahen, K. A., O., P. P., Pavasovic, A., Jenkins, R. E., and Strohbehn, K. (1991). Current-mode subthreshold MOS circuits for analog VLSI neural systems. *IEEE Transactions on Neural Networks*, 2(2):205–213.

Andreou, A. G., Meitzler, R. C., Strohbehn, K., and Boahen, K. A. (1995). Analog VLSI neuromorphic image acquisition and pre-processing systems. *Neural Networks*, 8(7/8):1323–1347.

Arreguit, X. (1989). *Compatible Lateral Bipolar Transistors in CMOS Technology: Model and Applications*. PhD thesis, Ecole Polytechnique Federale de Lausanne.

Bachelder, I. A., Waxman, A. M., and Seibert, M. (1993). A neural system for mobile robot visual place learning and recognition. *Proceedings of the World Congress on Neural Networks (WCNN'93)*, pages 512–517.

Baloch, A. A. and Waxman, A. M. (1991). Visual learning, adaptive expectations, and behavioral conditioning of the mobile robot MAVIN. *Neural Networks*, 4:271–302.

Baraldi, A. and Parmiggiani, F. (1995). A neural network for unsupervised categorization of multivalued input patterns, an application of satellite image clustering. *IEEE Transactions on Geoscience and Remote Sensing*, 33:305–316.

Bastos, J., Steyaert, M., Pergoot, A., and Sansen, W. (1997). Mismatch characterization of submicron MOS transistors. *Analog Integrated Circuits and Signal Processing*, 12:95–106.

Bastos, J., Steyaert, M., Roovers, R., Kinger, P., Sansen, W., Graindourze, B., Pergoot, A., and Janssens, E. (1995). Mismatch characterization of small size

MOS transistors. *Proc. IEEE 1995 Int. Conf. Microelectronic Test Structures*, 8:271–276.

Bernardon, A. M. and Carrick, J. E. (1995). A neural system for automatic target learning and recognition applied to bare and camouflaged SAR targets. *Neural Networks*, 8:1103–1108.

Bezdec, J. C. and Pal, S. K. (1992). *Fuzzy Models for Pattern Recognition*. IEEE Press.

Bezdek, J. (1981). *Pattern Recognition with Fuzzy Objective Function Algorithms*. Plenum, New York.

Bult, K. and Geelen, G. J. G. M. (1991). The CMOS gain–boosting technique. *Analog Integrated Circuits and Signal Processing*, 1:99–135.

Bult, K. and Wallinga, H. (1987). A class of analog CMOS circuits based on the square–law characteristic of a MOS transistor in saturation. *IEEE Journal of Solid-State Circuits*, SC–22:357–365.

Carpenter, G. A. (1997). Distributed learning, recognition, and prediction by ART and ARTMAP neural networks. *Neural Networks*, 10(8):1473–1494.

Carpenter, G. A. and Gjaja, M. N. (1994). Fuzzy-ART choice functions. *Proceedings of the 1994 World Congress on Neural Networks (WCNN'94)*, 1:713–722.

Carpenter, G. A. and Grossberg, S. (1987). A massively parallel architecture for a self–organizing neural pattern recognition machine. *Computer Vision, Graphics, and Image Processing*, 37:54–115.

Carpenter, G. A. and Grossberg, S. (1991a). *Pattern Recognition by Self Organizing Neural Networks*. MIT Press, Cambridge, MA.

Carpenter, G. A., Grossberg, S., Markuzon, N., and Reynolds, J. H. (1992). Fuzzy-ARTMAP: A neural network architecture for incremental supervised learning of analog multidimensional maps. *IEEE Transactions on Neural Networks*, 3:698–713.

Carpenter, G. A., Grossberg, S., and Reynolds, J. H. (1991b). ARTMAP: Supervised real–time learning and classification of nonstationary data by a self–organizing neural network. *Neural Networks*, 4:759–771.

Carpenter, G. A., Grossberg, S., and Rosen, D. B. (1991c). Fuzzy ART: Fast stable learning and categorization of analog patterns by an adaptive resonance system. *Neural Networks*, 4:759–771.

Caudell, T. P. and Healy, M. J. (1994a). Adaptive resonance theory networks in the encephalon autonomous vision system. *Proceedings of the 1994 IEEE International Conference on Neural Networks (ICNN'94)*, pages 1235–1240.

Caudell, T. P., Smith, S. D. G., Escobedo, S. D. G., and Anderson, M. (1994b). NIRS: Large scale ART1 neural architectures for engineering design retrieval. *Neural Networks*, 7:1339–1350.

Cauwenberghs, G. (1995a). A micropower CMOS algorithmic A/D/A converter. *IEEE Transactions on Circuits and Systems, Part I*, 42(11):913–919.

Cauwenberghs, G. (1997). Analog VLSI stochastic perturbative learning architectures. *International Journal of Analog Integrated Circuits and Signal Processing*, 13(1/2):195–209.

Cauwenberghs, g. and Pedroni, V. (1995b). A charge-based CMOS parallel analog vector quantizer. In Touretzky, D. S., editor, *Advances in Neural Information Processing Systems*, pages 779–786. MIT Press, Cambridge, MA.

Cauwenberghs, G. and Yariv, A. (1992). Fault-tolerant dynamic multi-level storage in analog VLSI. *IEEE Transactions on Circuits and Systems, Part II*, 41(12):827–829.

Choi, J. and Sheu, B. J. (1993). A high–precision VLSI winner–take–all circuit for self organizing neural networks. *IEEE Journal of Solid–State Circuits*, 28.

Christodoulou, C. G., Huang, J., Georgiopoulos, M., and Liou, J. J. (1995). Design of gratings and frequency selective surfaces using Fuzzy ARTMAP neural networks. *Journal of Electromagnetic Waves and Applications*, 9:17–36.

Chu, L. C. (1991). Fault–tolerant model of neural computing. *IEEE Int. Conf. on Computer Design: VLSI in Computers and Processors*, pages 122–125.

Cohen, M., Abshire, P., and Cauwenberghs, G. (1998). Mixed-mode VLSI implementation of fuzzy ART. *Proceedings of the 1998 IEEE International Symposium on Circuits and systems (ISCAS'98)*.

Degrauwe, M. G., Rijmenants, J., Vittoz, E. A., and De Man, H. J. (1982). Adaptive biasing CMOS amplifiers. *IEEE Journal of Solid–State Circuits*, SC–17:522–528.

Dubes, R. and Jain, A. (1988). *Algorithms that Cluster Data*. Prentice Hall, Englewood Cliffs, NJ.

Dubrawski, A. and Crowley, J. L. (1994). Learning locomotion reflexes: A self-supervised neural system for a mobile robot. *Robotics and Autonomous Systems*, 12:133–142.

Duda, R. and Hart, P. (1973). *Pattern Classification and Scene Analysis*. Wiley, New York.

Elias, S. A. and Grossberg, S. (1975). Pattern formation, contrast control, and oscillations in the short term memory of shunting on–center off–surround networks. *Biological Cybernetics*, 20:69–98.

Fahlman, S. E. (1989). Faster-learning variations on back–propagation: An empirical study. In Touretzky, D. S., Hinton, H., and Sejnowski, T., editors, *Proc. 1988 Connectionist Summer School*, pages 38–51. Morgan–Kaufmann, San Mateo, CA.

Gan, K. W. and Lua, K. T. (1992). Chinese character classification using adaptive resonance network. *Pattern Recognition*, 25:877–888.

Gersho, A. and Gray, R. M. (1992). *Vector Quantization and Signal Compression*. Kluwer, Norwell, MA.

Gilbert, B. (1975). Translinear circuits: A proposed classification. *Electronics Letters*, 11(1):14–16.

Gilbert, B. (1990a). Current–mode circuits from a translinear viewpoint: A tutorial in analog IC design: The current–mode approach. *IEE Circuits and Systems Series 2*, 2:11–91.

Gilbert, B. (1990b). Current-mode circuits from a translinear viewpoint, a tutorial. In Toumazou, C., Lidgey, F. J., and G., H. D., editors, *In Analogue IC Design: the current-mode approach*, pages 11–91. IEE Circuits and Systems Series 2.

Gjerdingen, R. O. (1990). Categorization of musical patterns by self-organizing neuronlike networks. *Music Perception*, 7:339–370.

Gopal, S., Sklarew, D. M., and Lambin, E. (1994). Fuzzy neural networks in multi-temporal classification of landcover change in the sahel. *Proceedings of the DOSES Workshop on New Tools for Spatial Analysis*, pages 55–68.

Grasmann, U. (1997). Process and apparatus for monitoring a vehicle interior. *US Patent number 5,680,096*.

Gregor, R. W. (1992). On the relationship between topography and transistor matching in an analog CMOS technology. *IEEE Transactions on Electron Devices*, 39(2):275–282.

Grossberg, S. (1976). Adaptive pattern classification and universal recoding I: Parallel development and coding of neural feature detectors. *Biological Cybernetics*, 23:121–134.

Grossberg, S. (1980). How does the brain build a cognitive code? *Psychological Review*, 87:151.

Ham, F. M. and Cohen, G. M. (1996a). Determination of concentrations of biological substances using raman spectroscopy and artificial neural network discriminator. *US Patent number 5,553,616*.

Ham, F. M. and Han, S. (1996b). Classification of cardiac arrhythmias using Fuzzy ARTMAP. *IEEE Transactions on Biomedical Engineering*, 43(4):425–430.

Hartigan, J. (1975). *Clustering Algorithms*. Wiley, New York.

Haykin, S. (1994). *Neural Networks: A Comprehensive Foundation*. IEEE Press and Macmillan College Publishing Co.

He, Y., Ciringiroglu, U., and Sánchez-Sinencio, E. (1993). A high–density and low–power charge–based hamming network. *IEEE Transactions on VLSI Systems*, 1:56–62.

Himmelbauer, W., Furth, P. M., Pouliquen, P. O., and Andreou, A. G. (1996). Log-domain filters in subthreshold MOS. Technical Report 96-03, The Johns Hopkins University, Electrical and Computer Engineering Dept.

Hochet, B., Peiris, V., Abdot, S., and Declercq, M. J. (1991). Implementation of a learning kohonen neuron based on a new multilevel storage technique. *IEEE Journal of Solid State Circuits*, 26:262–267.

Jones, S., Sammut, K., Nielsen, C., and Staunstrup, J. (1991). Toroidal neural network: Architecture and processor granularity issues. In Ramacher, U. and Rueckert, U., editors, *VLSI Design of Neural Networks*, pages 229–254. Kluwer Academic Publishers, Dordrecht, Netherlands.

Kalkunte, S. S., Kumar, J. M., and Patnaik, L. M. (1992). A neural network approach for high resolution fault diagnosis in digital circuits. *Proceedings of the International Joint Conference on Neural Networks*, pages 83–88.

Kasperkiewicz, J., Racz, J., and Dubrawski, A. (1995). HPC strength prediction using artificial neural network. *Journal of Computing in Civil Engineering*, 9:279–284.

Keulen, E., Colak, S., Withagen, H., and Hegt, H. (1994). Neural network hardware performance criteria. *Proceedings of the 1994 Int. Conf. on Neural Networks (ICNN'94)*, pages 1885–1888.

Kim, J. H., Lursinsap, C., and Park, S. (1991). Fault–tolerant artificial neural networks. *Int. Joint Conf. on Neural Networks (IJCNN'91)*, 2:951.

Kim, J. W., Jung, K. C., Kim, S. K., and Kim, H. J. (1995). Shape classification of on-line chinese character strokes using ART1 neural network. *Proceedings of the World Congress on Neural Networks (WCNN'95)*, pages 191–194.

Koch, M. W., Moya, M. M., Hostetler, L. D., and Fogler, R. J. (1995). Cueing, feature discovery, and one-class learning for synthetic aperture radar automatic target recognition. *Neural Networks*, 8:1081–1102.

Kohonen, T. (1989). *Self–Organization and Associative Memory*. Springer–Verlag, Berlin, Germany.

Kosko, B. (1987). Adaptive bidirectional associative memories. *Applied Optics*, 26:4947–4960.

Kumar, N., Himmelbauer, W., Cauwenberghs, G., and Andreou, A. G. (1997). An analog VLSI chip with asynchronous interface for auditory feature extraction. *Proceedings of the 1997 IEEE International Conference on Circuits and Systems*, 1:553–556.

Lakshmikumar, K. R., Hadaway, R. A., and Copeland, M. A. (1986). Characterization and modeling of mismatch in MOS transistors for precision analog design. *IEEE Journal of Solid-State Circuits*, SC-21(6):1057-1066.

Lang, K. J. and Wittbrock, M. J. (1989). Learning to tell two spirals apart. In Touretzky, D. S., Hinton, H., and Sejnowski, T., editors, *Proc. 1988 Connectionist Summer School*, pages 52-59. Morgan-Kaufmann, San Mateo, CA.

Lazzaro, J., Ryckebush, R., Mahowald, M. A., and Mead, C. (1989). Winner-take-all networks of O(n) complexity. In Touretzky, D. S., editor, *Advances in Neural Information Processing Systems*, volume 1, pages 703-711. Morgan-Kaufmann, Los Altos, CA.

Lin, D., Dicarlo, L. A., and Jenkins, J. M. (1988). Identification of ventricular tachycardia using intracavitary electrograms: analysis of time and frequency domain patterns. *Pacing and Clinical Electrophysiology*, 11:1592-1606.

Linares-Barranco, B., Sánchez-Sinencio, E., Rodríguez-Vázquez, A., and Huertas, J. L. (1992). A modular T-mode design approach for analog neural network hardware implementations. *IEEE Journal of Solid-State Circuits*, 27(5):701-713.

Linares-Barranco, B., Sánchez-Sinencio, E., Rodríguez-Vázquez, A., and Huertas, J. L. (1993). A CMOS analog adaptive BAM with on-chip learning and weight refreshing. *IEEE Transactions on Neural Networks*, 4(3):445-455.

Lubkin, J. and Cauwenberghs, G. (1998). A micropower learning vector quantizer for parallel analog-to-digital data compression. *Proceedings of the 1998 IEEE International Symposium on Circuits and systems (ISCAS'98)*.

Ly, S. and Choi, J. J. (1994). Drill conditioning monitoring using ART1. *Proceedings of the IEEE International Conference on Neural Networks (ICNN'94)*, pages 1226-1229.

Lyon, R. F. and Schediwy, R. R. (1987). CMOS static memory with a new four-transistor memory cell. In Losleben, P., editor, *Advanced Research in VLSI*, pages 110-132. MIT Press, Cambridge, MA.

Mauduit, N., Duranton, M., Gobert, J., and Sirat, J. A. (1992). Lneuro 1.0: A piece of hardware lego for building neural network systems. *IEEE Transactions on Neural Networks*, 3(3):414-422.

Mead, C. (1989). *Analog VLSI and Neural Systems*. Addison Wesley, Reading, MA.

Medeiro, F., Rodríguez-Macías, R., Fernández, F. V., Domínguez-Castro, R., and Rodríguez-Vázquez, A. (1994). Global analogue cell design and optimization using statistical techniques. *Analog Integrated Circuits and Signal Processing*, 6:179-195.

Mehta, B. V., Vij, L., and Rabelo, L. C. (1993). Prediction of secondary structures of proteins using Fuzzy ARTMAP. *Proceedings of the World Congress on Neural Networks (WCNN'93)*, pages 228–232.

Michael, C. and Ismail, M. (1992). Statistical modeling of device mismatch for analog MOS integrated circuits. *IEEE Journal of Solid–State Circuits*, 27(2):154–166.

Moore, B. (1989). ART1 and pattern clustering. In Touretzky, D. S., Hinton, H., and Sejnowski, T., editors, *Proc. 1988 Connectionist Summer School*, pages 174–185. Morgan–Kaufmann, San Mateo, CA.

Murshed, N. A., Bortozzi, F., and Sabourin, R. (1995). Off-line signature verification, without a priori knowledge of class $\omega 2$. A new approach. *Proceedings of the Third International Conference on Document Analysis and Recognition (ICDAR'95)*.

Nairn, D. G. and Salama, C. A. T. (1990a). Current–mode algorithmic analog–to–digital converters. *IEEE Journal of Solid–State Circuits*, 25:997–1004.

Nairn, D. G. and Salama, C. A. T. (1990b). A ratio–independent algorithmic analog–to–digital converter combining current mode and dynamic techniques. *IEEE Transactions on Circuits and Systems*, 37:319–325.

Neti, C., Schneider, M. H., and Young, E. D. (1992). Maximally fault tolerant neural networks. *IEEE Transactions on Neural Networks*, 3(1):14–23.

Pan, T. W. and Abidi, A. A. (1989). A 50-dB variable gain amplifier using parasitic bipolar transistors in CMOS. *IEEE Journal of Solid-State Circuits*, 24(4):951–961.

Pao, Y. H. (1989). *Adaptive Recognition and Neural Networks*. Addison Wesley, Reading, MA.

Pelgrom, M. J. M., Duinmaijer, A. C. J., and Welbers, A. P. G. (1989). Matching properties of MOS transistors. *IEEE Journal of Solid–State Circuits*, 24:1433–1440.

Pouliquen, P. O., Andreou, A. G., and Strohbehn, K. (1997). Winner-takes-all associative memory: A hamming distance vector quantizer. *Journal of Analog Integrated Circuits and Signal Processing*, 13(1/2):211–222.

Rabiner, L. and Juang, B. H. (1993). *Fundamentals of Speech Recognition*. Prentice Hall, Englewood Cliffs, NJ.

Racz, J. and Dubrawski, A. (1995). Artificial neural network for mobile robot topological localization. *Robotics and Autonomous Systems*, 16:73–80.

Rade, L. and Westergren, B. (1990). *BETA Mathematics Handbook*. CRC Press.

Ramacher, U., Beichter, J., Raab, W., Anlauf, J., Bruels, N., Hachmann, U., and Wesseling, M. (1991). Design of a 1st generation neurocomputer. In Ramacher, U. and Rueckert, U., editors, *VLSI Design of Neural Networks*, pages 271–310. Kluwer Academic Publishers, Dordrecht, Netherlands.

Riedmiller, M. (1994). Advanced supervised learning in multi–layer perceptrons - from backpropagation to adaptive learning algorithms. *Int. Journal of Computer Standards and Interfaces*, 5.

Rodríguez-Vázquez, A., Domínguez-Castro, R., Medeiro, F., and Delgado-Resti-tuto, M. (1995). High resolution CMOS current comparators: Design and application to current–mode function generation. *International Journal of Analog Integrated Circuits and Signal Processing*, 7:149–165.

Rodríguez-Vázquez, A., Espejo, S., Domínguez-Castro, R., Huertas, J. L., and Sánchez-Sinencio, E. (1993). Current–mode techniques for the implementation of continous- and discrete–time cellular neural networks. *IEEE Transactions on Circuits and Systems II*, 40(3):132–146.

Roulston, D. J. (1990). *Bipolar Semiconductor Devices*. McGraw–Hill, New York.

Sackinger, D. and Guggenbuhl, W. (1990). A high–swing, high–impedance MOS cascode circuit. *IEEE Journal of Solid–State Circuits*, 25:289–298.

Sánchez-Sinencio, E. and Lau, C. (1992). *Artificial Neural Networks: Paradigms, Applications, and Hardware Implementations*. IEEE Press.

Sánchez-Sinencio, E., Ramírez-Angulo, J., Linares-Barranco, B., and Rodríguez-Vázquez, A. (1989). Operational transconductance amplifier-based nonlinear function syntheses. *IEEE Journal of Solid–State Circuits*, 24:1576–1586.

Schlimmer, J. S. (1987). *Concept Adquisition through Representational Adjustment (Technical Report 87–19)*. PhD thesis, Department of Information and Computer Sciences, University of California at Irvine.

Seevinck, E. and Wiegerink, J. (1991). Generalized translinear circuit principle. *IEEE Journal of Solid–State Circuits*, SC–26(8):1198–1102.

Seibert, M. and Waxman, A. M. (1992). Adaptive 3D object recognition from multiple views. *IEEE Transactions on Pattern Analysis and Machine Intelligence*, 14:107–124.

Seibert, M. and Waxman, A. M. (1993). An approach to face recognition using saliency maps and caricatures. *Proceedings of the World Congress on Neural Networks (WCNN'93)*, pages 661–664.

Serrano-Gotarredona, T. and Linares-Barranco, B. (1994). The active–input regulated–cascode current mirror. *IEEE Transactions on Circuits and Systems, Part I: Fundamental Theory and Applications*, 41(6):464–467.

Serrano-Gotarredona, T. and Linares-Barranco, B. (1995). A modular current–mode high–precision winner–take–all circuit. *IEEE Transactions on Circuits and Systems, Part II*, 42(2):132–134.

Serrano-Gotarredona, T. and Linares-Barranco, B. (1996a). A modified ART1 algorithm more suitable for VLSI implementations. *Neural Networks*, 9(6):1025–1043.

Serrano-Gotarredona, T. and Linares-Barranco, B. (1996b). A real–time clustering microchip neural engine. *IEEE Transactions on Very Large Scale Integration (VLSI) Systems*, 4(2):195–209.

Serrano-Gotarredona, T. and Linares-Barranco, B. (1997). An ART1 microchip and its use in multi-ART1 systems. *IEEE Transactions on Neural Networks*, 8(5):1184–1194.

Serrano-Gotarredona, T. and Linares-Barranco, B. (1998a). A high–precision current–mode WTA–MAX circuit with multi–chip capability. *IEEE Journal of Solid–State Circuits*, 33(2):280–286.

Serrano-Gotarredona, T., Linares-Barranco, B., and Andreou, A. G. (1998b). Voltage clamping current mirrors with 13-decades gain adjustment range suitable for low power MOS/bipolar current mode signal processing circuits. *Proceedings of the 1998 IEEE International Symposium on Circuits and Systems (ISCAS'98)*.

Sheingold, D. H. (1976). *Nonlinear Circuits Handbook*. Analog Devices Inc., Norwood, MA.

Shyu, J. B., Temes, G. C., and Krummenacher, F. (1984). Random error effects in matched MOS capacitors and current sources. *IEEE Journal of Solid–State Circuits*, SC–19(6):948–955.

Soliz, P. and Donohoe, G. W. (1996). Adaptive resonance theory neural network for fundus image segmentation. *Proceedings of the 1996 World Congress on Neural Networks (WCNN'96)*, pages 1180–1183.

Srinivasa, N. and Sharma, R. (1996). A self-organizing invertible map for active vision applications. *Proceedings of the World Congress on Neural Networks (WCNN'96)*, pages 121–124.

Strader, N. R. and Harden, J. C. (1989). Architectural yield optimization. In Schwartzlander, E. E., editor, *Wafer Scale Integration*, pages 57–118. Kluwer Academic Publishers, Boston.

Tarng, Y. S., Li, T. C., and Chen, M. C. (1994). Tool failure monitoring for drilling processes. *Proceedings of the Third International Conference on Fuzzy Logic, Neural Nets and Soft Computing, Iizuka, Japan*, pages 109–111.

Tse, P. and Wang, D. D. (1996). A hybrid neural networks based machine condition forecaster and classifier by using multiple vibration parameters. *Proceedings of the 1994 IEEE International Conference on Neural Networks*, pages 2096–2100.

Tsividis, Y. (1988). *Operation and Modeling of the MOS Transistor*. McGraw–Hill, New York.

Wand, Y. F., Cruz, J. B., and Mulligan, J. H. (1990). On multiple training for bidirectional associative memory. *IEEE Transactions on Neural Networks*, 1:275–276.

Waxman, A. M., Seibert, M. C., Gove, A., Fay, D. A., Bernardon, A. M., Lazott, C., Steele, W. R., and Cunningham, R. K. (1995). Neural processing of targets in visible, multispectral IR and SAR imagery. *Neural Networks*, 8:1029–1051.

Wienke, D. (1994). Neural resonance and adaptation - towards nature's principles in artificial pattern recognition. In Buydens, L. and Melssen, W., editors, *Chemometrics: Exploring and Exploting Chemical Information*. University Press, Nijmegen, NL.

Yuille, A. L. and Geiger, D. (1995). Winner–take–all mechanisms. In Arbib., M. A., editor, *The Handbook of Brain Theory and Neural Networks*, pages 1056–1060. MIT Press.

Zadeh, L. (1965). Fuzzy sets. *Information and Control*, 8:338–353.

Index

Active-input, 97, 100, 124
Back propagation, 36, 55-56
Bipolar transistor, 43, 122-124, 126, 158-159, 161, 164
Cardiac, 168, 172-173
Cascode, 60, 97-98, 100, 109, 140
  regulated, 62, 97-100, 109
Category proliferation, 20, 26, 29, 31
Channel length modulation, 97
Charge
  injection, 150, 160-161
  pump, 150, 160
Choice function, 3, 24, 39-40, 42, 51, 57, 117, 128, 130-131, 135-137, 146, 149, 155-157, 159, 167
Class, 1-2, 14-16, 18-20, 23, 31, 141-142, 179
Coarseness, 10, 192
Committed/uncommitted node, 6-7, 11-12, 14, 26-27, 47, 50-51, 60, 141, 193, 200, 202-206
Comparator, 3, 161-162
  current, 58-59, 62, 64, 68, 70, 94-97, 107, 140-141, 149
Complement coding, 20-21, 26-27, 29, 32-33, 145, 147, 149, 155, 171, 174
Complexity, 90-92
Decay, 147, 151-153, 164
Device electrical parameters, 211
Device physical parameters, 210-211
Direct access, 12-13, 157, 192-195, 199
Distance, 3, 42, 45, 47, 49, 154, 157-159
  hamming, 45
Division operation, 39, 41-43, 137, 139
Don't know, 20, 142, 172, 179
Early voltage, 123, 159
Electrocardiogram, 172-173
Encoder decoder, 66, 130-131, 141
Fast commit slow-recode, 26
Fast learning, 4, 23, 36-38, 55, 191-192

Fault, 68, 74, 76, 80, 141, 215-216
Feature, 7, 9-11, 21, 26, 41, 173-174
Floorplan, 128-129
Fuzzy
  max, 157, 159
  min, 24, 118, 126, 145-147, 149, 157, 159, 162, 206
  subset, 206-207
  superset, 207
Glucose, 169, 171
Gradient, 77-80, 159, 211
Hspice, 99, 105, 114
Impedance, 97-100, 107, 109-110, 130, 212
Interconnections, 56
  per second, 55, 71
  update per second, 56, 71
Layout, 63, 65-66, 68, 122-124, 126, 128-129, 142-143, 159-160, 211
Leakage, 145, 147, 151-152, 158
Least mean squares, 216
Levenberg-marquardt, 214
Linear predictive coding, 173
Map-field, 14, 16, 18, 21-22, 31, 36, 38
Match-tracking, 16, 18, 20-23
Matlab, 2, 7, 21, 30, 34, 36, 177
Memory, 21, 23, 40, 44, 56, 59, 120-122, 126, 128, 158, 165
  LTM, 4, 6, 191-192, 197, 199
  STM, 4, 191-192, 196
Mismatch, 64-66, 68-69, 77, 79-80, 96
  random, 57, 60, 65, 74, 77, 86, 97, 99-100, 109
  systematic, 57, 63, 97
Modular, 63, 149
Normalize, 7, 26-27, 29, 32
On/off-line, 12, 20, 23, 38, 53-56, 174-175
Oscillation, 105
Ota, 62, 100
Pole, 105, 124-125

Qrs complex, 173–174
Ram, 130–131, 141–142
Rare events, 12, 23, 26
Refresh, 145, 148, 150, 152–156, 158–162, 164
Self
  organizing, 23, 158, 164
  refresh, 165
  scaling, 10, 192
  stabilization, 11, 84, 199
Shift register, 67, 74, 76, 131, 141, 174
Sigmoid, 102, 105, 109
Stability, 23, 92, 100, 105, 107–109, 124, 151–152, 163–165
Stable category learning, 29, 198
Standard deviation, 47, 65, 69, 78–80, 96, 99, 210, 213, 215, 222
Subset/superset, 12–13, 20, 157, 192–193, 199–207
Switch, 60, 63, 66, 70, 132, 160–162

Translinear, 137, 139, 149–150
Tree mirror, 65–66, 79–80
Trip point voltage, 62, 102
Two spiral problem, 31
Vector quantizer, 146, 154, 162
Vigilance, 10, 145, 149, 151–152, 159
  circuit, 135–136, 140
  criterion, 6–7, 12, 16, 21, 24–26, 44, 50–51, 54, 59, 70, 73, 131, 145–146, 149, 152, 155, 157–158, 167
  parameter, 4, 6, 10–11, 14, 16, 18, 20, 23, 28, 30, 36, 38, 44, 73, 82, 85–86, 146, 149, 151, 192, 206
  subsystem, 3–6, 10–11, 13–14, 23, 25, 192, 195–198, 200–201
Vlsi-friendly, 39, 41, 54, 57, 155–157
Voting strategy, 36, 38
Weak inversion, 43, 117, 126, 137, 139–140
Yield, 68, 74, 76, 80